GEOLOGICAL FLUID DYNAMICS

Sub-surface Flow and Reactions

Owen Phillips's textbook, *Flow and Reactions in Permeable Rocks*, published in 1991, became a classic in the field of geological fluid dynamics. This book is its long-awaited successor. In the intervening years, significant advances have been made in our understanding of subterranean flow, especially through the vast amount of research into underground storage of nuclear waste and aquifer pollution. This new book integrates and extends these modern ideas and techniques and applies them to the physics and chemistry of sub-surface flows in water-saturated, sandy, and rocky media. It describes essential scientific concepts and tools for hydrologists and public health ecologists concerned with present-day flow and transport, and also for geologists who interpret present-day patterns of mineralization in terms of fluid flow in the distant past. The book is ideal for graduate students and professionals in hydrology, water resources, and aqueous geochemistry.

OWEN M. PHILLIPS is a Fellow of the Royal Society and a member of the US National Academy of Engineering. He has held academic posts at Cambridge University and the Johns Hopkins University. He was awarded the Sverdrup Gold Medal of the American Meteorological Society for his contributions to oceanography, and a fellowship in the American Geophysical Union for his contributions to geological fluid dynamics. His *Last Chance Energy Book*, published in 1979, anticipated the first global energy crisis of the 1980s, while his recent research has been on sub-surface aquifer flows, the dispersal of contaminants and flow-controlled reactions in rocks. He has two other publications with Cambridge University Press – *The Dynamics of the Upper Ocean* (1966), which was awarded the Adams Prize from Cambridge University, and *Flow and Reactions in Permeable Rocks* (1991).

GEOLOGICAL FLUID DYNAMICS

Sub-surface Flow and Reactions

O. M. PHILLIPS, FRS

Decker Professor Emeritus at the Johns Hopkins University

CAMBRIDGE
UNIVERSITY PRESS

University Printing House, Cambridge CB2 8BS, United Kingdom

One Liberty Plaza, 20th Floor, New York, NY 10006, USA

477 Williamstown Road, Port Melbourne, VIC 3207, Australia

314-321, 3rd Floor, Plot 3, Splendor Forum, Jasola District Centre, New Delhi - 110025, India

79 Anson Road, #06-04/06, Singapore 079906

Cambridge University Press is part of the University of Cambridge.

It furthers the University's mission by disseminating knowledge in the pursuit of education, learning and research at the highest international levels of excellence.

www.cambridge.org
Information on this title: www.cambridge.org/9781108462068

© O. M. Phillips 2009

First published 2009
First paperback edition 2018

A catalogue record for this publication is available from the British Library

ISBN 978-0-521-86555-5 Hardback
ISBN 978-1-108-46206-8 Paperback

For Merle

Contents

Preface

This book is concerned with the dynamics of subterranean flows in the natural environment, with the transport and dispersal of contaminants that they may carry and the chemistry of the interactions between the matrix and the fluid that percolates through it in spatially random conduits or aquifer pores. It is intended for anyone with a quantitative interest in the world around them, and particularly for professionals and graduate students in hydrology, geology and environmental science and engineering. Many of the basic concepts originated in the late nineteenth or mid twentieth centuries, but they were usually developed in a very idealized and simplified form because few field measurements of actual sub-surface seepage or flow patterns had been attempted. Only recently have extensive and detailed hydrological field measurements been undertaken and their findings contain surprises that are re-defining the way we view this part of the natural world and begin to understand how it works.

In writing this book, I have relied heavily on the guidance and advice of many colleagues, friends and students who listened patiently, corrected gently and pointed in new directions. In particular, I must thank Lawrence Hardy, John Ferry, Jim Wood and Gordon Wolman, all colleagues, Robert Shedlock of the US Geological Survey and Emory Cleaves of the Maryland Geological Survey and my Cambridge colleagues, Andrew Woods and Herbert Huppert who always had something new to show me. I am particularly grateful to my wife for her tireless reading of the manuscript and her suggestions for improvement.

O. M. P.
Chestertown

1

Introduction

The relatively new scientific field of Geological Fluid Mechanics is concerned with applying the principles of fluid mechanics to the geological sciences. It is characterized by close interaction between carefully conceived laboratory measurements on geological flows and theoretical analyses that interpret the results in terms of basic physical principles. It was given this name by Herbert Huppert in Cambridge, one of its leading practitioners in a company that includes Andrew Woods, also in Cambridge, George Veronis at Yale, Stewart Turner at ANU, Canberra, and many other technically powerful and imaginative scientists. This present book concentrates on the part of Geological Fluid Mechanics that involves the flow of passive and reacting fluids through porous or fractured geological media. In our planet, both hot and cold aqueous fluids have flowed or seeped through sand and fractured rocks for eons, modifying their composition by dissolution, chemical reaction and deposition. Great crystalline formations and mineral deposits were formed by nature during that time and modifications continue naturally. The study of these processes was always an interesting intellectual challenge, but one of no particular urgency. Yet within a couple of lifetimes the pace of change has exploded as a result of human activity.

The contamination of our aquifers, and in turn the rivers and estuaries into which the groundwater flows, is the result of both deliberate and inadvertent injection of a variety of human, agricultural and industrial wastes, but our knowledge of the extent of these changes is meager. How long does it take for the contaminants to build up, where do they go and, if we remove the source, how long will it take for the contamination to flush out? What happens to the effluents from coal mines and paper mills, that are dumped into streams? Nuclear power plants help to satisfy our gluttonous appetite for energy without generating the primary greenhouse gas, carbon dioxide, but the bargain is Faustian – high-level, long-lived radioactive wastes are being stored on site at the nuclear power plants that generated them.

The long-term disposal of these wastes requires that they be removed from human contact for times much longer than all of human history.

Geological fluid dynamics is concerned with how these natural systems work, with their patterns of flow and chemical reaction in a variety of geological media, whether sandy aquifers, layered sediments or mosaics of fractured rocks. The first geological fluid dynamicist was probably Henry Darcy (1803–58), although he certainly did not think of himself as such. A very accomplished hydraulic engineer, he worked on the Dijon water supply for a number of years, and towards the end of his life, he and two assistants conducted a series of hydraulic experiments on the flow of water through a vertical column partially filled with siliceous sand from the Saone River and a flow of water from the Dijon hospital water supply. The volume flux of water was measured in a gauging station, the pressure difference across the sand bed was measured by using two mercury U-tube manometers, and he found a very accurate linear relation between the two. His work was published as an appendix to his extensive report (Darcy, 1856) on the public fountains of the city of Dijon.

Darcy's study exemplifies the three essential ingredients in the scientific exploration of the nature of this world about us. The explorer needs to have (i) a detailed and soundly based understanding of the basic rules, the "laws of nature" under which the flows operate, (ii) a continuing contact with physical reality and familiarity with the results of whatever careful experiments, observations and detailed measurements that have been made on these flows, and (iii) the ability to put the two together. Darcy obviously knew a lot about hydraulics, he performed the experiment himself and he made the critical quantitative connection between flow rate and pressure gradient.

In more recent times and on a larger scale we can discern these same three attributes that have guided the remarkable progress during the past 50 years of our sister science, Meteorology. The "laws of nature" that govern the motion and properties of the atmosphere are essentially the same as those outlined in the next chapter of this book, the conservation laws of thermodynamics, of Newtonian continuum dynamics, of chemical reactions, etc., supplemented in the atmosphere by the laws of radiation. The "physical reality" is the atmosphere itself, in constant motion and burdened by its increasing load of carbon dioxide and other greenhouse gases. Measurements of atmospheric pressure and temperature have been made for over 200 years, but the pace increased in the 1970s when the Global Atmospheric Research Program (GARP) stimulated a vast increase in the systematic observation, measurement and monitoring of the atmosphere that continues today. This in turn provided a strong stimulus to the development of "super-computers" that were needed to handle the new floods of data and the numerical models of large-scale atmospheric motions being developed by Jules Charney and others. New techniques were developed at the National Center for Atmospheric Research in Boulder, Colorado, for airborne measurements while remote sensing systems such as the atmospheric radar of David Atlas at NASA were becoming able to scan for clouds, rain and atmospheric motions over larger and larger volumes of the atmosphere. Today, on the evening TV weather report we see marvelous real-time, data-based computer simulations of local and regional weather that are vivid, ongoing and generally accurate. We receive warnings of rain tomorrow and impending dangers such as hurricanes a thousand miles away.

In contrast, the quantitative base of data in the geological sciences is much more sparse. It is probably safe to assert that more atmospheric measurements are made every day, than have been made on geological flows in all of recorded history. Subterranean flow measurements are difficult and expensive to acquire by drilling and the data are sometimes classified for commercial reasons. The medium is solid, often hard, opaque, and complex in structure and composition. Seismic techniques are able to delineate internal structures, but most of the information that we have still comes from surface exposures and patterns of seepage. As a result of all these factors, the quantitative measurements that we do have are extremely valuable but still severely limited in number and scope. Particularly notable are the recent measurement programs on the dispersal of tagged fluids in aquifers, conducted mainly by the US and state Geological Surveys and their analogs in Europe. These have generated a leap forward in our understanding of the structure of the variations in permeability in sandy substrates and the spatially random flow field that percolates through them.

The relative paucity of field data on geological flows presents a mis-match with the power and sophistication of modern digital computers. With few exceptions, numerical simulations of geological flows have little measured data input, or quantitative comparison between the computer output and field measurements. Parameters can be chosen without observational or experimental basis, but simply to make the output "seem reasonable," i.e. to be in accord with preconceptions. Though often presented as factual, and generating their own air of reality, these simulations are often quite misleading, and no more than digitally precise renditions of a mostly imaginary world. There is little doubt that a more fruitful approach would include the development of relatively simple models with several essential ingredients: (i) the powerful but often neglected physical constraints such as minimum dissipation, (ii) the use of measured parameters, (iii) the pertinent physical and chemical balances involved and (iv) the flexibility for application to a variety of possible structural configurations. The results must then be evaluated critically by comparison with whatever laboratory or field data that does exist.

In the next chapter of this book, the general geometrical characteristics of permeable media are described, together with the basic physical and chemical balances that underlie the developments that follow. Two important general theorems concerning uniqueness and minimum dissipation, dating from the nineteenth century and often overlooked, provide useful insights into the structure of constant density flows in complex regions with variable matrix permeability. The first part of Chapter 3 is a brief summary of some "classical" porous media flows, the basic concepts of groundwater age and the various time scales for aquifer flows. Recent measurements have shown that the spatially random, horizontally isotropic permeability structure of a nominally uniform sandy aquifer is associated with highly *anisotropic* dispersive characteristics of dissolved contaminants. Dissolved solutes in fracture–matrix media disperse rapidly in the longitudinal direction, but much more slowly in transverse directions. These findings can be understood best in terms of the minimum dissipation constraint. Unsteady flows are also of interest. Pressure pulses from explosions and seismic eruptions spread rapidly at acoustic speeds, but the residual pressure then relaxes diffusively as interstitial gas and liquids present flow out of fractured porous rock in seeps or geysers, at a rate that diminishes in time.

Chapter 4 describes the nature of buoyancy-driven flows from convection plumes to freshwater wedges and gravity currents. These flows are qualitatively different from uniform-density flows and characteristically possess circulation in the transport velocity field. In a given geological structure, the flow patterns are no longer unique, which raises the possibility of instability and spontaneous evolution of one flow pattern into another. The archetypical thermal instability is associated with the name of Rayleigh (1916) and, since then, many variations of the basic theme have been discovered that depend on the different rates of diffusion of heat and dissolved salts in permeable media. These instability processes have been found in laboratory measurements, in contemporary natural flows and their traces left in ancient rocks.

Chapter 5 synthesizes these flow patterns with the patterns of reaction, deposition and dissolution that the flow produces when the interstitial fluids and the matrix interact chemically. There are three dominant flow-mediated reaction scenarios, reaction fronts, gradient reactions and mixing zones, each of which has characteristic patterns of occurrence. In many geological scenarios, the rates of reaction may be limited by the rate at which the flow can deliver dissolved solutes to the reaction site. When dissolved contaminants in a surface aquifer are absorbed into, or react with, the enclosing matrix, a patch of contaminant moves considerably more slowly than does the interstitial fluid, centimeters per day, perhaps. An extreme situation is found when the reaction involves replacement and the solutions are dilute compared with the mass per unit volume of the solid reactant. The propagation speed of a reaction front may be smaller than the interstitial fluid velocity by *many* orders of magnitude, possibly being only a few centimeters per millennium. Application of these concepts and results can illuminate not only the formation of mineral deposits in paleogeologic time, but also also the accumulation, transport and dispersal of dissolved contaminants in present-day aquifers.

2

The basic principles

2.1 Pores and fractures

The geological materials with which we are concerned usually lie at one extreme of the range of "porous media" encountered in nature and technology. The porosity, the volume fraction of connected voids that allow fluid movement, may be as large as 0.3 or 0.4 in a well-sorted sandbank or as small as 1% in natural calcite (Pryor, 1973). The skeletal remains of corals that abound in tropical reefs contain myriad interstices on scales of up to a centimeter or so and may have a similarly large porosity; Figure 2.1 shows a sample from Bermuda at approximately half-scale. This kind of structure containing fluid conduits as well as more numerous smaller pores is at the high-porosity extreme of those generally encountered. Compaction by overlying sediments, the infilling of interstices by finer grains, and the precipitation of cements from solution can reduce the porosity by an order of magnitude and reduce the permeability, as we shall see, by three orders of magnitude or more.

Many large pores are also apparent in dolomite from the Latemar Massif in northern Italy (Figure 2.2). Calcium ions from the original calcite mineral have been replaced by magnesium, generating dolomite. The specific volume of the dolomite produced in the reaction is less by 3–13% than that of the original calcite, so that as the reaction proceeded, the porosity increased.

Networks of small cracks or fractures allow fluid percolation even when the matrix itself is relatively impermeable. Seepage from fractures can often be discerned in roadside rock exposures. Figure 2.3 illustrates a smaller-scale network in a sandstone cleavage plane, made visible by stain. Stained fluid moves relatively rapidly through the fracture network but spreads laterally into the matrix blocks only slowly. Note the progression of wider, older stains passing through the whole sample, from which spring shorter, narrower and newer branches.

Fault systems also provide conduits for fluid motion. Even a cursory glance at many field exposures often reveals layers of quartz or other minerals apparently

6

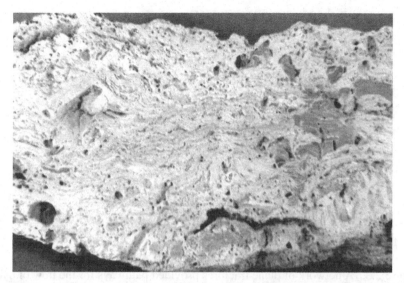

Figure 2.1. An extremely porous limestone from a Bermuda coral reef, approximately half-scale, containing shell fragments and many interstices with scales of up to a centimeter or so, courtesy of Professor L. Hardie.

Figure 2.2. Pores in dolomite from the Latemar Massif in northern Italy. The blocks in the scale are 1 cm long. (Photograph courtesy of Dr. E. N. Wilson.)

deposited along fractures in a larger matrix. Figure 2.4 shows a mosaic of small scale fractures that have served as pathways for fluid motion until becoming filled by deposition from the infiltrating solution, with subsequent fractures appearing later.

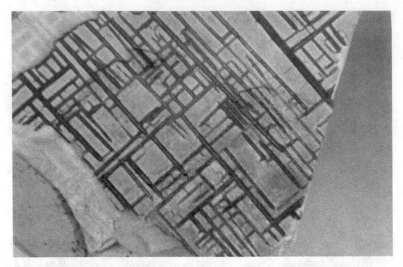

Figure 2.3. A network of plane fractures provided pathways for the flow of dyed fluid in a sandstone cleavage plane, which then diffused into the matrix blocks, courtesy of Professor L. Hardie. Approximately full scale.

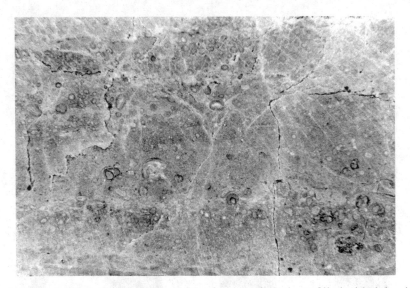

Figure 2.4. In this calcite block, previous fractures have been filled with dolomite, while more recent fractures remain partially open.

2.2 Geometrical characteristics

2.2.1 Porosity

A number of geometrical length scales are pertinent to flow through permeable rocks and aquifer matrices. In aqueous solutions, the intermolecular or inter-ionic distances are of order 10^{-9} m, while the scale of the smallest interstices of hydrological interest is possibly 1000 times larger, i.e. 1 micron or more. As a result, the fluid can be regarded as a continuum whose motion through the passages of the medium can be described in terms of the concepts and equations of continuum fluid mechanics. The photographs of the previous section indicate that in porous limestone or partly cemented sandstone, the characteristic diameter δ_0 of the orifices or interstices may be as large as 10^{-4}–10^{-3} m, with a relative few even larger. In sand, the distance between the individual flow paths, the diameter of the interstices and the grain size are all comparable. In more consolidated rocks and granites, however, the size of the interstices is characteristically much smaller than the microscale distance l_0 between them. Media may also be extensively fractured on scales from 10^{-2} to 10 m (see Figure 2.3) and such fracture networks are potentially important flow conduits. Sedimentary deposits are frequently bedded, with local variations in physical properties such as the porosity and permeability occurring over vertical scales that are large compared with the grain size but small compared with the overall thickness of the bed. Finally, there are the macroscopic length scales h, which specify the thickness (or smallest dimension) of the porous bed as a whole, and l, its lateral extent.

In order to relate the overall flow behavior to the average geometrical characteristics of the rocks, we must consider carefully certain statistical aspects of the microscopic flow through individual pores or regions of inhomogeneity, as in Section 2.10 below. Our primary concerns are with flow patterns and velocities, with the transports of heat and chemical solutes on the scale of structural variations or on the macroscopic length scale h of the structure itself. Immense simplifications are possible when the microscale l_0 is sufficiently small compared with h that we can find an intermediate local scale, the matrix averaging scale l_{AV}, that is large compared with the grain or matrix block size l_0 yet small compared with the scale h of the flow patterns that we wish to resolve. Thus we require that

$$\delta_0 \leq l_0 \ll l_{AV} \ll h, \tag{2.1}$$

where, as a rule of thumb, the $\ll$ inequality signs can be interpreted to mean "is less by a factor of at least 10 than." Within a given stratum, properties of the medium and characteristics of the flow through it, when averaged over the volume l_{AV}^3 are expected to vary smoothly from one averaging volume to the next. When the system is viewed on a macroscopic scale, it can again be regarded as a continuum, with the "point" properties being, in fact, local averages of this kind, functions of three spatial coordinates (x, y, z) and possibly time t.

When the locally averaged properties of the medium are independent of the position of the averaging volume, the medium is said to be homogeneous. If all locally averaged properties are independent of direction, the medium is described as locally isotropic. A well-mixed sand bed may be, on this averaging scale, both homogeneous and locally isotropic. As mentioned previously, however, sediments are frequently deposited in such a way that the fabric preserves a record in its layering of the vertical direction, though there may be no differences discernible in the two orthogonal horizontal directions. Seasonal variations in sedimentation rate may produce, on a microscopic scale, a stack of horizontal laminations that, on a macroscopic scale, lead to different percolation characteristics along and across the laminations. Similarly, when elongated or plate-shaped sedimenting particles tend to settle horizontally, the resulting fabric will preserve a record of the vertical direction at the time of deposition. In general, when the medium has one preferred direction but its properties are independent of rotation about that direction, it can be described as locally axi-symmetrical. In spite of these caveats, a uniform sandbank does provide the basic prototype of a classic hydrodynamical porous medium, whose essential geometry, involving a three-dimensional web of minute intersecting fluid pathways, is found in many porous rocks and other geologic media at different scales and with different detailed topologies. It is convenient to call these "sandbank-type" media to distinguish them from the fracture–matrix media described later, which obey the same basic conservation laws but whose geometry gives them quite different flow characteristics.

An important characteristic of a porous rock is the void fraction. The total void fraction ϕ_T is that fraction of the total averaging volume represented by the interstices; the solids occupy a fraction $1 - \phi_T$ of the whole. This can be measured by an examination of randomly taken thin sections; since a volumetric sample can be considered to be a stack of plane slices, the ratio of void area to total area in a typical thin section is equal to the ratio of void volume to total volume, ϕ_T. Similarly, along a sampling line the average ratio of total length of the line segments in voids to total line length is also ϕ_T. However, not all of the void spaces may be active in fluid flow through the medium. Isolated cavities or "dead end" tubes can contribute to ϕ_T but do not provide microscopic pathways to fluid motion. A more significant measure for our purposes is the connected porosity ϕ, in which only those voids that provide connections among the averaging volumes are considered. In general, this quantity cannot be estimated from thin-section examination without some additional information or assumption about the structure of the fabric, but it can be measured by comparison of the mass of saturated and dried samples or, as we shall see in the next section, by fluid observations. In this book, the term "porosity" refers to the connected porosity, since this is the property of interest in fluid motion, though in the petroleum industry it is commonly used as a synonym for "void fraction."

Clearly, $\phi \leq \phi_T < 1$. At one extreme, in a material with the geometry of Swiss cheese, such as pumice, all the voids are isolated so that $\phi = 0$ while ϕ_T may be relatively large. In poorly cemented sandstone, there may be very few isolated voids and $\phi \cong \phi_T \sim 0.2$ or 0.3. In consolidated rocks, $\phi \sim 0.02$–0.1, while in an assembly of well-sorted grains, the porosity may be as large as 0.3 or 0.4.

A variety of textures are of interest in different contexts, and some simple approximate relationships are frequently useful. If, on a microscopic scale, the interstices consist of a complex of intersecting convoluted pathways, the porosity can be represented as the number of pathways per unit area found in a thin section, n, times the average area of their intersections with the thin section, which is proportional to δ_0^2. Thus,

$$\phi \sim n\delta_0^2. \tag{2.2}$$

2.2.2 Double porosity in a fracture–matrix medium

Not all hydrological flow media have the statistical geometry of a sandbank. Fractures on many scales are ubiquitous in rock strata and even casual observation of roadway cuttings after wet weather show them to provide pathways of significant fluid flow. A fracture *plane* (the usual terminology) should not be considered as an approximately uniform gap between two parallel rock surfaces, but as a more-or-less planar network of intersecting, ribbon-shaped fluid pathways around areas of close rock contact and possible accumulations of detritus. Let λ represent the total area of the ribbon pathways in the fracture plane per unit volume of the medium. By again visualizing the volume as a stack of slices, it can be seen that λ also approximates the total length of the intersections of the ribbon pathway network per unit area in any slice through the medium. If the *mean* width of the fluid pathways is $\bar{\delta}_F$, the fracture porosity, the fraction of the total volume that they occupy is

$$\phi_F = \lambda \bar{\delta}_F. \tag{2.3}$$

Although $\bar{\delta}_f$ is the appropriate measure of the fracture aperture for specifying the fracture porosity, it is shown in Section 2.4 that it is far less relevant to the magnitude of the mean fracture flow. For a given local pressure gradient, the flow velocity depends on a high power of the gap width, and consequently upon the distribution of fracture apertures; the fluid volume flux (velocity times gap width) is disproportionately larger through the wider gaps.

It is interesting to compare typical numerical values of the fracture porosity ϕ_F with values of the matrix porosity in the blocks between the fractures. In a moderately consolidated porous rock, $\phi \sim 0.2$, while for the individual pathways, δ_0 may be 10^{-6}–10^{-5} m, so that from (2.2), the number n of internal pathways intersecting unit area is very large. Fractures produced by mechanical failure of the matrix may have an average gap $\overline{\delta}_F$ of 10^{-5} or 10^{-4} m, and with a moderate density of pathway intersections, $\lambda \sim 3\,\text{m}^{-1}$ (length per unit area), the fracture porosity is only about 10^{-4}, which is smaller by a factor of order 2000 than that of the matrix blocks. Clearly, the block porosity provides the dominant reservoir for interstitial fluid, though as will be found later, the cracks can provide the dominant pathways for flow.

2.3 The transport velocity and mass conservation

Consider the fluid flow in a sandbank-type medium in which the mean pore velocity is represented by $\overline{v}$. The porosity or active void fraction is ϕ, so that over the fraction ϕ of a unit matrix volume, the mean flow velocity is $\overline{v}$ while over the rest, the solids or inactive voids, it is zero. Consequently, averaged over the whole, the volume flux or the fluid volume flow per unit cross-sectional area or the transport velocity

$$\mathbf{u} = \phi\overline{\mathbf{v}}. \tag{2.4}$$

The velocity $\mathbf{u}$ can be interpreted as the velocity with which a fluid *would be moving* if it occupied the whole space and had the same volume transport. Its direction is parallel to $\overline{v}$, but since for many geological materials ϕ is numerically small, $\mathbf{u}$ is substantially smaller in magnitude than $\overline{v}$.

The porosity of a rock sample is frequently estimated by comparing wet and dry weights of a sample. It could also be measured directly by taking advantage of equation (2.4), especially if the identification of active and inactive voids is difficult. If water is forced under pressure through the sample, the volume of water passing through per unit cross-sectional area per unit time is the transport velocity. If a spot of dynamically passive dye or other marker (i.e. one that does not affect the flow) is introduced, it will pass through the sample on average at the mean interstitial speed, and the ratio of the two defines ϕ.

The transport velocity $\mathbf{u}$ is one of the primary field variables used in this book, since transports, not simply of fluid volume but also of heat and chemical species, are of primary interest. There are two important things to remember about it. First, if we are concerned with the spreading of *particular* elements of fluid that can be marked with salinity, dye, or other passive contaminants, then we are necessarily concerned with the *interstitial* or pore velocity, not the transport velocity; this has some profound consequences, as we shall see later. Secondly, note that because of the particular averaging definition above, the transport velocity does not obey the usual rule of vector addition. If the medium is moving, being subducted, say, at velocity $\mathbf{V}$, the interstitial fluid is indeed moving at velocity $\overline{v} + \mathbf{V}$, but only through the fraction ϕ of a transverse plane, so that the transport velocity relative to a fixed reference frame is $\phi(\bar{\mathbf{v}} + \mathbf{V})$, or $\mathbf{u} + \phi\mathbf{V}$.

2.3.1 Mass conservation

One physical constraint on the flow patterns through porous fabrics that can be expressed simply in terms of $\mathbf{u}$ is the conservation of total fluid mass. In a fixed arbitrary finite region of the fabric, the mass of interstitial fluid is

$$\int \phi\rho dV,$$

where ϕ is the porosity, ρ is the interstitial fluid density, and the integral is throughout the volume of the region. This fluid mass may change as a result of net fluid flow across the bounding surface, the net mass transport outward being

$$\int \rho\mathbf{u} \cdot d\mathbf{S},$$

where the integral is over the surface of the region and the element of surface area $d\mathbf{S} = \mathbf{n}d\mathbf{S}$ is directed outward. It may also change if water is generated internally at the rate ρS_W per unit fabric volume by chemical dehydration reactions among the rock constituents. The net rate of change of fluid mass in the region is therefore

$$\frac{\partial}{\partial t}\int \phi\rho dV = -\int \rho\mathbf{u}.d\mathbf{S} + \int \rho S_\mathrm{W}dV$$
$$= \int \{-\nabla \cdot (\rho\mathbf{u}) + \rho S_\mathrm{W}\}dV,$$

by the divergence theorem. Since the volume is taken as fixed, the initial time derivative can be taken inside the integral:

$$\int \{\partial(\phi\rho)/\partial t + \nabla \cdot (\rho\mathbf{u}) - \rho S_\mathrm{W}\}dV = 0, \tag{2.5}$$

and since the volume is also arbitrary, the integrand itself must vanish. A local statement of fluid mass conservation is then

$$\frac{\partial}{\partial t}(\phi\rho) + \nabla \cdot (\rho\mathbf{u}) = \rho S_\mathrm{W}. \tag{2.6}$$

In this statement, the rate of change of interstitial water mass in the medium, reduced by the divergence of the mass transport velocity field, is equal to the rate of generation of water by rock reactions, represented by the final source term.

2.3.2 The incompressibility condition

For many purposes, the complete version of (2.6) is unnecessarily general. It can be rewritten in the somewhat disassembled form,

$$\nabla \cdot \mathbf{u} = S_W - \frac{\partial \phi}{\partial t} - \left(\phi \frac{\partial}{\partial t} + \mathbf{u} \cdot \nabla \right) \ln (\rho / \rho_0), \tag{2.7}$$

that sorts out those physical effects included in the general form (2.6) that are usually less important. The first term on the right of (2.7), the source term, is significant only when water is being *generated* by chemical reaction or absorbed into the matrix. The next expresses the simple geometrical fact that if the porosity decreases with time, $(\partial \phi / \partial t < 0)$, there is a flow divergence as fluid is expelled from the decreasing fraction of voids. This can be important in tectonic events, but if the porosity of the rock changes over geological time as cementation or dissolution occurs, the time scales are long, and although the cumulative changes in ϕ may be large, the *rates* of change are extremely small. During and following seismic events, significant compaction may occur in clays or other unconsolidated sediments and ϕ may change rapidly. It has sometimes been suggested that compaction is an important process in driving interstitial fluid flow, but when such sediments compact, the total volume of fluid expelled per unit volume of matrix is just the change in porosity, necessarily a small fraction of unity. Natural geological fluids are generally dilute and significant geochemical changes require movement through the matrix of fluid volumes that are many times the matrix volume itself. The interstitial fluid velocities associated with compaction are negligible compared with those produced by maintained hydraulic or thermal forcing.

Variations in space and time of the density factors in the last term of equation (2.7) are also usually negligible. Variations in fluid density occurring as a result of temperature variations may be very important in producing variations in buoyancy of the interstitial fluids which give rise to convective motion, but since the *fractional* change in fluid density $\delta \rho / \rho_0$ is very small, its influence on the flow divergence is minor. In geological flow fields, the individual terms in $\nabla \cdot \mathbf{u}$ are of order u/l, where u is the transport speed and l the overall flow dimension; the last term is smaller by order $\delta \rho / \rho_0$ and is therefore negligible unless boiling occurs. Another exceptional circumstance arises when the interstitial fluid pressure is suddenly altered – for example, by a natural earthquake or artificially in a pressurized well. The interstitial fluid is often much more compressible than the rock matrix, and if the pressure is suddenly reduced by a nearby fracture, the un-fractured rock porosity changes little but the interstitial fluid will seek to expand, forcing a time-dependent, local flow, a process Sibson calls "seismic pumping," but one that may be more accurately described as pressure diffusion, considered in Section 3.6 of this book.

Subject to these provisos, the terms involving variations in ϕ and ρ will be ignored unless they are centrally involved in the application at hand, and in the absence of dehydration reactions as well, equation (2.6) simplifies to

$$\nabla \cdot \mathbf{u} = 0. \tag{2.8}$$

This equation expresses the statement that the volumetric divergence of the fluid vanishes, and is usually called the incompressibility condition. In Cartesian coordinates, with $\mathbf{u} = (u, v, w)$,

$$\nabla \cdot \mathbf{u} = \frac{\partial u}{\partial x} + \frac{\partial v}{\partial y} + \frac{\partial w}{\partial z} = 0. \tag{2.9}$$

An integral form of this statement is frequently useful. In a fixed region of arbitrary shape, the volume integral of the divergence of the velocity field is, by the divergence theorem, equal to the normal efflux over the bounding surface. Thus, from (2.8),

$$\int \nabla \cdot \mathbf{u} \, dV = \int \mathbf{u} \cdot dS = 0, \tag{2.10}$$

and the net volume flux into or out of the region vanishes.

A complementary conceptual approach is to view the conservation of fluid mass in terms of individual fluid elements. As argued above, we can usually assume that the porosity ϕ is constant in time. The balance (2.6) can be written alternatively in terms of the mean interstitial velocity $\bar{\mathbf{v}}$ of (2.4) as

$$\frac{d\rho}{dt} = \frac{\partial \rho}{\partial t} + \bar{\mathbf{v}} \cdot \nabla \rho = -\rho \nabla \cdot \bar{\mathbf{v}} + \phi^{-1} \rho S_{\mathrm{w}}. \tag{2.11}$$

The Lagrangian operator $d/dt = \partial/\partial t + \bar{\mathbf{v}} \cdot \nabla$ in this equation is used widely in considerations of fluid flow. The first part, $\partial/\partial t$, represents the rate of change in time at a fixed spatial point, while $\bar{\mathbf{v}} \cdot \nabla$ represents the rate of change that is observed at a point moving with velocity $\bar{\mathbf{v}}$ up the gradient of a *spatially* variable property. The combination, written as d/dt, is then the rate of change observed in a the moving frame of reference when the property itself varies both in space and time, and is called "the derivative following the motion." It is formally equivalent to the total derivative in multivariate calculus, but its physical interpretation in this context is useful and important. In (2.11), the rate of change of the fluid density following the interstitial motion is equal to minus the divergence (i.e. the convergence) of fluid mass plus the rate of generation of fluid mass from dehydration reactions, if any. If the fluid density does not change following the motion, the left-hand side of (2.11) vanishes and if there are no dehydration reactions, the last term also is zero. So $\nabla \cdot \bar{\mathbf{v}} = 0$ and since the porosity ϕ is constant, we recover (2.8).

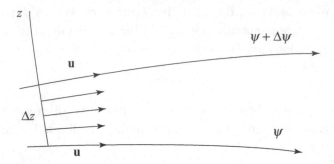

Figure 2.5. In two-dimensional flow, the fluid transport between neighboring streamlines ψ and $\psi + \Delta\psi$, the fluid transport $u\,\Delta z = \Delta\psi$ is constant, so that as their spacing increases, the fluid velocity decreases.

2.3.3 The stream function

In the special case of incompressible, two-dimensional flow, the stream function is a quantity that provides a graphic and quantitatively accurate image of the flow, and has the additional advantage in analysis of reducing the number of flow variables by one. In two-dimensional flow, the incompressibility condition (2.8) becomes

$$\frac{\partial u}{\partial x} + \frac{\partial v}{\partial y} = 0, \tag{2.12}$$

which is always satisfied by a function $\psi(x, y, t)$ such that

$$u = \frac{\partial \psi}{\partial y}, \qquad v = -\frac{\partial \psi}{\partial x}. \tag{2.13}$$

The function ψ is called the stream function. Clearly, if we choose a differentiable but otherwise arbitrary function ψ, then (2.13) specifies a kinematically possible flow field, i.e. one that satisfies the two-dimensional incompressibility condition (2.12). The inverse statement is also true. The statement (2.12) is the condition that $u\delta x - v\delta y$ is an exact differential, $\delta\psi$, say, from which (2.13) follows.

The velocity component in any direction is thus represented as the gradient of the stream function ψ in the orthogonal direction; in particular, in the direction in which $\psi = $ const., there is *no* transverse velocity, so that the contours $\psi = $ const. represent the directions of flow at each point. Moreover, if we choose local coordinates with the origin at a point P, and with the x-axis chosen to lie along the flow direction as in Figure 2.5, then at the neighboring point Q, a distance Δz from P, the stream function is

$$\psi + \Delta\psi = \psi + \frac{\partial \psi}{\partial z}\Delta z,$$
$$= \psi + u\,\Delta z.$$

The difference in the values of the stream function at the two points is then $\Delta \psi = u \Delta z$, the volume flux across the line joining the two points. Streamline patterns contain a great deal of useful and accurate information. In a distribution of contours, for each of which $\psi = $ const., the volume flux between any pair of these curves remains constant. Where the streamlines are close together, the volume transports are concentrated and the flow speed is relatively high; where they are widely separated, the flow speed is low. Streamlines cannot begin or end in the interior of a region. In re-circulating flow, the streamlines are closed; if fluid enters or leaves a porous region across a bounding surface, streamlines originate or terminate at that surface, whereas at an impermeable boundary (with no flow across the surface) the stream function is constant along it. Figure 3.5 gives an example of the use of the stream function for the representation of flow in a simple surface aquifer with distributed infiltration from rainfall across the water table.

Note that the numerical value of the stream function is arbitrary to the extent of an additive constant, since only *differences* in its value at different points have physical significance. The value of the stream function can be assigned (usually zero) for one particular streamline; the values of ψ along the other streamlines then give the total volume transport between that streamline and the streamline $\psi = 0$. In *steady* two-dimensional flow, the streamlines correspond to the transport paths of marked fluid elements (except for diffusion), but in unsteady flow this is not generally so.

In mathematical terms, use of the stream function enables us to represent the two velocity components $u(x, y, t)$, $v(x, y, t)$ in terms of the single scalar function $\psi(x, y, t)$ while satisfying the incompressibility condition automatically. We then have one variable fewer and one equation fewer – any such simplification is always welcome!

A different stream function, the Stokes stream function, can be defined for axially symmetrical flow, which has this same mathematical advantage but a slightly different interpretation in terms of the flow pattern. In cylindrical polar coordinates (r, θ, z), with corresponding velocity components (u, v, w), the incompressibility condition is

$$\nabla \cdot \mathbf{u} = \frac{1}{r} \frac{\partial}{\partial r}(ru) + \frac{1}{r} \frac{\partial v}{\partial \theta} + \frac{\partial w}{\partial z} = 0. \tag{2.14}$$

When the flow is axially symmetrical, $\partial / \partial \theta = 0$, and this reduces to

$$\frac{1}{r} \frac{\partial}{\partial r}(ru) + \frac{\partial w}{\partial z} = 0, \tag{2.15}$$

which is satisfied by the function ψ_S, the Stokes stream function, such that

$$u = \frac{1}{r} \frac{\partial \psi_S}{\partial z}, \qquad w = -\frac{1}{r} \frac{\partial \psi_S}{\partial r}. \tag{2.16}$$

As before, lines $\psi_S = $ const. indicate the direction of flow, so that their pattern gives a good visual representation of the flow field. The total volume transport between the axially symmetrical *surfaces* $\psi_S = \psi_1$ and ψ_2 is $2\pi(\psi_1 - \psi_2)$, but the r^{-1} factors in (2.16) produce a geometrical distortion in the relationship between fluid velocities and the gradients of ψ_S at different radial positions.

2.4 Darcy's law

2.4.1 Hydrostatics

In a fluid at rest, the pressure increases with depth to support the weight of the overlying fluid; the pressure gradient is vertically downward. For a change dz in depth, the pressure changes by the amount $\rho_0 g \, dz$, where ρ_0 is the local fluid density and g the gravitational acceleration. The hydrostatic pressure gradient can therefore be expressed as

$$\nabla p_h = (0, 0, -\rho_0 g) = -\rho_0 g \mathbf{l}, \tag{2.17}$$

where $\mathbf{l}$ is a unit vector vertically upward. This remains true no matter what the shape of the container enclosing the fluid and no matter whether the fluid as a whole is over- or under-pressurized. In the present context, (2.17) holds true for *connected regions of single phase* fluids at rest inside the interstices of permeable rocks.

The water table is identified by the water level in an un-pumped well open to the atmosphere. It is defined in this book as the surface at which the interstitial water pressure is equal to the mean atmospheric pressure. If the groundwater is not moving, the water table is horizontal. It does not generally coincide with the upper interface of the water-saturated region, since surface tension can draw water upward until the capillary suction from air–water interfaces, concave on the air-side, supports the weight of water above the level of the water table. J. R. Philip has made many important contributions on water movement in soils; specific references are contained in his review (Philip, 1989). In rocks, liquid water may be immobilized by some degree of chemical bonding with the minerals in the matrix. Because of the randomness in the pore geometry of most permeable rocks, the interface between the air and the liquid water region can be expected to be highly irregular with isolated pockets of water and air in the unsaturated regions above and below. Its geometry is self-adjusting and over short time intervals it is presumably essentially static, like a raft of bubbles on water. For a clean air–water interface, the surface tension coefficient T_C is about 73 dynes/cm (or ergs/cm^2) at 15 °C, though in field conditions, dissolved surface-active and biological materials may reduce the effective surface considerably. The capillary suction, the difference between the air pressure and the water pressure immediately below an interface is given by Laplace's formula,

$$\delta p = T_C \left(\frac{1}{R_1} + \frac{1}{R_2} \right), \tag{2.18}$$

where R_1 and R_2 are the principal radii of curvature, which in this context are of the order of half the pore size. The finer are the pores, the greater is the capillary suction and the thicker is this capillary zone. In an extremely fine matrix, its maximum possible height above the water table may be limited by cavitation when the absolute pressure of the liquid drops below the vapor pressure at that temperature. In the laboratory under very clean and static conditions, cavitation in water can be avoided over a significant range of negative absolute pressures, but this would not be expected in natural rocks with an abundant supply of nucleation sites such as sharp corners, specks of debris, etc.

2.4.2 Interstitial flow through a uniform matrix

If the interstitial pressure gradient ∇p is not hydrostatic and the weight of fluid per unit volume ρg is not necessarily uniform, the fluid will flow at a rate determined by $-\nabla p - \rho g \mathbf{l}$, the driving force per unit volume. The signs express the facts that fluid flow is driven *down* the pressure gradient and that the weight of fluid ρg acts downwards while the unit vector is conventionally taken as vertically upward. Once the interstitial fluid is in motion, internal viscous stresses are generated that oppose the motion, and the resulting fluid velocity in the pores is determined by the balance the driving and the resistive forces. In small-molecule fluids such as water or aqueous solutions, the viscous stresses (force per unit area) are proportional to the rates of strain in the fluid, the constant of proportionality being the molecular viscosity μ, a property of the fluid. Fluids in which the stress, rate-of-strain relation is linear are called Newtonian fluids. In the interstices, the rates of strain depend on the local geometry but are of order v/δ, where v represents the interstitial fluid velocity and δ the characteristic pore size, so that the viscous stresses are proportional to $\mu v/\delta$. In an individual fluid pathway, the retarding viscous force acting along the walls (per unit length of path) is this stress times the pore circumference, which is of order δ, i.e. μv. The driving force is $-(\nabla p + \rho g \mathbf{l})$ times the pore area of order δ^2 over which it acts. The balance between them can be expressed in terms of the *mean* interstitial velocity $\bar{\mathbf{v}}$ as

$$\mu \bar{\mathbf{v}} \propto -\delta^2 (\nabla p + \rho g \mathbf{l}). \tag{2.19}$$

In terms of the transport velocity $\mathbf{u} = \phi\bar{\mathbf{v}}$,

$$\mathbf{u} = -\frac{k}{\mu}(\nabla p + \rho g\mathbf{l}), \tag{2.20}$$

where k is called the intrinsic permeability. It is proportional to $\phi\delta^2$ with a numerical constant of proportionality that incorporates the complicated statistical geometry of the connected pore spaces in the medium. This relation is known as Darcy's law.

An important comment should be made at this point. The linearity of the relation found by Darcy (1856) between the driving forces and the transport velocity is a consequence of both the linear viscosity relation in a Newtonian fluid (such as air, water or aqueous solutions) and also the neglect of inertial effects in the pore fluid as it moves along its convoluted pathways. If, at some point, its trajectory has a radius of curvature r, the fluid inertia sets up an additional pressure gradient $\rho v^2/r$, where v is the pore velocity – this provides the centripetal acceleration associated with the curved trajectory. Darcy's law is accurate only when these inertial pressure gradients are small compared with the viscous stress gradients $\mu v/\delta^2$, and since in general $r \sim \delta$, this requires that

$$\frac{\rho v^2}{\delta} \ll \frac{\mu v}{\delta^2},$$

or that the combination

$$\frac{v\delta}{\nu} = \frac{u\delta}{\phi\nu} = R \ll 1, \tag{2.21}$$

where $\nu = \mu/\rho$ is called the kinematic viscosity and R is the pore Reynolds number which expresses the ratio of inertial to viscous effects in the flow. A corresponding inequality is assumed in fracture flow. This condition is usually satisfied very strongly in sub-surface flow except possibly for flow through coarse gravel beds or in vigorous geothermal systems. For such cases, Forchheimer represented the additional form drag by adding to the linear drag a term quadratic in the transport velocity and containing an empirical form drag coefficient. Other less fortunate embellishments of the Darcy equation include the subtraction of a term $\mu\nabla^2\mathbf{u}$ from the pressure gradient in (2.20), the result being known as the Brinkman equation (see Nield, 1984) and of the addition of a cubic drag term, giving what has been called the cubic Forchheimer equation. Both of these, however, seem to be of more mathematical than hydrological interest.

There are, however, situations in which the simple Darcy equation is inadequate. If the interstitial fluid consists of two or more separate components or phases, such as oil and water, or air and water, the radius of curvature of the interface can be comparable with the pore size δ. From Laplace's formula (2.18), surface tension T_C can support a pressure difference of order T_C/δ across the interface without motion. If n represents the number of such interfaces per unit length along the pores, an overall pressure gradient of order nT_C/δ can be supported without fluid motion. When this threshold pressure gradient is exceeded, fluids will begin to move, first in the pathways with largest pores and, subsequently, as the pressure gradient increases, in the pathways with the next largest and so on. The *distribution* of pore sizes, and not simply the mean, becomes crucial in determining the flow characteristics, and the overall relation between pressure gradient and transport velocity is certainly nonlinear. Interesting numerical experiments on capillary displacement and percolation of immiscible fluids, showing fingering, trapping, and incomplete displacement of one fluid by the other, have been conducted by Chandler, Koplik, Lerman and Willemsen (1982), but this very important though specialized topic will not be pursued in this book.

2.4.3 Permeability

The permeability is an intrinsic property of the medium, just as the viscosity is an intrinsic property of the interstitial fluid. The permeability k has physical dimensions L^2 and is measured in units of cm^2 or m^2 and so forth. A curious unit, but one still widely used in hydrology, is the darcy, defined as the permeability that allows a transport velocity of 1 cm/s of a fluid with viscosity 1 centipoise (10^{-2} c.g.s. units, close to that of water) under a pressure gradient of 1 atmosphere per centimeter! Since mean atmospheric pressure is $\sim 10^6$ c.g.s. units, a permeability of 1 darcy is about 10^{-8} cm^2, or 10^{-12} m^2. A related quantity also widely used when pressures are expressed in terms of the equivalent hydraulic head is the hydraulic conductivity

$$K = \frac{\rho g k}{\mu} = \frac{g k}{\nu} \tag{2.22}$$

where $\nu = \mu/\rho$ is called the *kinematic* viscosity. The hydraulic conductivity depends on both the permeable medium and the fluid; since its physical dimensions are $[LT^{-1}]$ (as are those of velocity), it is expressed in units of cm/s, m/s, or m/yr, etc. It is a particularly convenient quantity in nearly horizontal groundwater flow in unconfined aquifers in which the pressure is close to hydrostatic (see below) and the Darcy equation reduces to the simple two-dimensional form

$$\mathbf{u} = -K\nabla_H\zeta, \tag{2.23}$$

where $\nabla_H = (\partial/\partial x, \partial/\partial y)$ is the horizontal gradient operator and $\zeta(x, y, t)$ is the water table configuration. In a water-saturated medium, the numerical value of K in m/s is about 10^7 larger than the numerical value of k in m^2.

The formulation of the Darcy equation (2.20) does not involve assumptions about any specific geometry of the fluid pathways in a three-dimensional medium, though the derivation above does assume that the pathway scale can be represented by the characteristic pore diameter δ alone. In the proportionality statement $k \propto \phi\delta^2$, the numerical values of the proportionality constant are found to be generally quite small. Some simple geometries can be solved explicitly. If the interstices form an ensemble of parallel tubes of diameter δ, the Poiseuille solution of 1840 (see, for example, Batchelor, 1967) leads to a permeability of $k = (\phi\delta^2/32)\cos\theta$, where θ is the angle between the axis of the tubes and the pressure gradient. With an isotropic distribution of tube directions, the permeability is $\phi\delta^2/96$, even smaller than the previous case because many of the tubes are transverse to the pressure gradient. In more complex geometries, the permeability cannot be calculated so simply (if at all!) but it is of interest to note that while the permeability in each case is proportional to $\phi\delta^2$, the constant of proportionality is small, in the range $(1-3) \times 10^{-2}$. If, in a particular sample, the porosity $\phi = 0.1$ and the pore size $\delta = 0.1$ mm, we would expect k to be in the range $(1-3) \times 10^{-11}$ m^2, or $(1-3) \times 10^{-7}$ cm^2. Most laboratory measurements of permeability or saturated hydraulic conductivity involve the measurement of volume flux through a specimen under a large imposed pressure gradient and the use of equation (2.20), or more recently, in a centrifuge where the centrifugal potential gradient provides the driving force (Nimmo and Mello, 1991).

Note also that the grain size *per se* is irrelevant; the porosity involves the size of the pores. The two are proportional only in the case of well-sorted, un-cemented grains, needles, or other particles of approximately uniform shape and size. In spite of this, attempts have been made to express various sets of empirical data on permeability in terms of porosity and grain size (see, for example, the summary in Lerman 1979), but the empirical formulas obtained should be used for media similar to those measured or, at most, only as a very general guide.

In a useful review (concerned mainly with crystalline rocks), Brace (1980) points out that laboratory measurements of permeability in small samples give values for sandstone of from 10^{-12} to 10^{-16} m^2 and for limestone–dolomite, a wide range from 10^{-14} to as small as 10^{-22} m^2, reflecting the variations in micro-structural characteristics. Metamorphic and granitic rocks have permeabilities from 10^{-16} to 10^{-21} m^2. In situ measurements give generally higher values, 10^{-9}–10^{-12} m^2 for chert and reef limestone, 10^{-12}–10^{-18} m^2 for granites or crystalline rocks, probably reflecting flow through fractures rather than through the rock itself. Stober (1996) reports similar values in Black Forest granites and gneisses. Fractures may be sealed or absent in shale. Many limestones and sandstones are already so permeable $(\sim 10^{-15}$ m$^2)$ that fractures add little. In the face of these variations, it is clearly rash to depend heavily on an *a priori* assumption or estimate of average values,

Table 2.1. *Characteristic magnitudes of permeability and porosity for some common rocks, orders of magnitude only. The higher values for in situ measurements probably result from flow through fractures, rather than through the matrix itself*

Material	Permeability (m^2)		Porosity
	Laboratory samples	In situ measurements	
Sandstone	$10^{-12}-10^{-16}$	$10^{-10}-10^{-14}$	5×10^{-2}
Limestone Dolomite	$10^{-14}-10^{-22}$	$10^{-10}-10^{-12}$	$(5-20) \times 10^{-2}$
Metamorphic/Granitic	$10^{-16}-10^{-21}$	$10^{-12}-10^{-16}$	10^{-2}

Data from Brace (1980) and Stober (1996).

but fortunately, as we shall see in subsequent chapters, flow *patterns* can often be inferred without such numerical guesses. Estimate of flow magnitudes, which are critical in many practical applications involving contaminant dispersion and extraction, do usually involve directly the numerical values of k and their spatial variability. Although field calibration is sometimes possible, this often remains a primary uncertainty (see Table 2.1).

2.4.4 Reduced pressure and buoyancy

In many groundwater flow situations, the pressure remains quite close to hydrostatic and in the Darcy equation (2.20), the two terms on the right are almost equal in magnitude but opposite in sign. Their sum (which drives the flow) is much smaller than the magnitude of either. It is often more accurate in calculations, and is also conceptually revealing in flows driven by heat or salinity, to subtract out from (2.20) the large term $\nabla p_H = -\rho_0 g \mathbf{l}$ representing the vertical pressure gradient in a fluid whose density ρ_0 is the average for the interstitial fluid in the domain of flow. The difference between the actual or total fluid pressure p_T and the hydrostatic pressure referred to a convenient origin for z is called the reduced pressure or the non-hydrostatic pressure. Since the total pressure is of little interest in flow considerations (though important in thermodynamics), the unadorned symbol p will henceforth refer to the reduced pressure $p_T - p_H$. With this understanding, after division by ρ_0, (2.20) becomes

$$\mathbf{u} = \frac{k}{\nu}(-\nabla(p/\rho_0) + b\mathbf{l}), \qquad (2.24)$$

where the kinematic viscosity $v = \mu/\rho_0$ and the combination

$$b = \left(\frac{\rho_0 - \rho}{\rho_0}\right) g \tag{2.25}$$

is the interstitial fluid buoyancy, a body force per unit mass that acts vertically upward where the local fluid density ρ is less than the mean ρ_0. Where $\rho > \rho_0$, the fluid has excess weight locally and the negative buoyancy is a body force per unit mass acting downward. Buoyancy has the same physical dimensions as acceleration (LT^{-2}). This form (2.24) of the Darcy equation exhibits clearly on its right-hand side the two major driving forces for sub-surface fluid motion. The first term is the gradient of reduced pressure, generally associated with variations in the water table level above, driving fluid flow down the gradient from higher to lower pressure. The second expresses the buoyancy of the interstitial fluid. If, because of variations in temperature or salinity in the medium (greater temperature or lower salinity), $\rho < \rho_0$, the fluid density is less than the mean and the buoyancy is positive, providing a body force upward.

Note particularly that the gradients in any horizontal direction of reduced pressure and total pressure are the same, since by definition the hydrostatic pressure gradient is vertical.

To specify the reduced pressure $p = p_T - p_H$ itself, rather than its gradient, we must take a reference level, say $z = 0$, for the hydrostatic pressure and, relative to this, the hydrostatic pressure $p_H = -\rho g z$, with z taken vertically upward. If the water table is at $z = \zeta(x, y)$, where the actual pressure p_T is atmospheric (i.e. zero), then the reduced pressure at the water table is

$$p = \rho g \zeta(x, y). \tag{2.26}$$

This is equivalent to the potential introduced by Hubbert (1940), a somewhat redundant concept that is used widely in the groundwater literature. Its usefulness is limited to constant density flows in which the velocity is proportional to the gradient of a scalar (the potential). In flows where buoyancy is important, the idea of a potential is no longer useful, but the concept of reduced pressure remains physically pertinent.

2.4.5 Boundary conditions

There are two general types of condition that must be satisfied at the boundaries of a flow domain in order to define the flow. So-called kinematical boundary conditions assert in general that fluid cannot be created nor disappear at a boundary. Dynamical boundary conditions assert that, except for capillarity, the fluid pressure is continuous across domain boundaries. A discontinuity in the pressure field would imply an infinite pressure gradient and infinite transport velocity, which is physically unacceptable. Thermal boundary conditions specify either the boundary temperature or heat flux, whichever is physically more appropriate. Some specific examples follow.

(1) There can be no fluid transport across a fixed impermeable boundary, so that the normal component of the transport velocity vanishes at the boundary. Symbolically, $\mathbf{u} \cdot \mathbf{n} = 0$ there. Streamlines cannot cross such a surface; in two dimensions the surface must coincide with a streamline $\Psi = \text{const}$. In the absence of buoyancy effects, Darcy's equation shows that the pressure gradient normal to the boundary also vanishes: $\partial p / \partial n = 0$.

(2) At a submerged boundary, between a saturated sub-strate and a lake above, the pressure is hydrostatic and the reduced pressure is the same as that at the water surface above.

(3) At internal boundaries between regions 1 and 2 of different permeability, say, the *normal* component of velocity is continuous across the interface; $\mathbf{u}_1 \cdot \mathbf{n} = \mathbf{u}_2 \cdot \mathbf{n}$, where the direction of the unit normal $\mathbf{n}$ is the same on each side. Consequently, from Darcy's law, $k_1 (\partial p / \partial n)_1 = k_2 (\partial p / \partial n)_2$ so that the normal pressure gradient on each side is inversely proportional to the permeability on that side.

(4) At all points along an internal interface, the pressure is the same on each side. Consequently the pressure gradient *along the interface* is the same on both sides, and from Darcy's law, the tangential velocity on each side is proportional to the permeability on that side. $\mathbf{u}_1 \cdot \mathbf{t}/k_1 = \mathbf{u}_2 \cdot \mathbf{t}/k_2$, where $\mathbf{t}$ is a tangential unit vector. If the permeability is discontinuous at an interface between two rock types, then so is the transport velocity along either side. Conditions (3) and (4) can be combined to produce an analogue of Snell's law in optics, $k \sin \theta = \text{const}$, where θ is the angle between the direction of flow and the normal.

However, if the boundary is "free," i.e. an interface between two identifiable bodies of fluid such as fresh and saline water, or air and water at the water table, it may move *through* the matrix. The dynamical boundary condition is again that the pressure is continuous across the moving interface, except for capillary effects. The appropriate kinematical boundary condition however, is slightly more complex. Consider the conservation of interstitial fluid at a free boundary $z = \zeta(x, y, t)$ with the z-coordinate vertically upward, as illustrated in Figure 2.6. If fluid is flowing beneath a fixed interface with slope $\nabla \zeta$, the vertical component of the fluid velocity is $\bar{\mathbf{v}}_H \cdot \nabla \zeta$, where $\bar{\mathbf{v}}_H$ is the horizontal component. If, in addition, the interface is moving upward with velocity $\dot{\zeta}$, the total vertical velocity $\bar{v}_Z$ of fluid at the interface is

$$\bar{v}_Z = \dot{\zeta} + \bar{\mathbf{v}}_H \cdot \nabla \zeta \qquad (2.27)$$

or, in terms of the transport velocities $(\mathbf{u}, w)$,

$$w = \phi \dot{\zeta} + \mathbf{u} \cdot \nabla \zeta. \qquad (2.28)$$

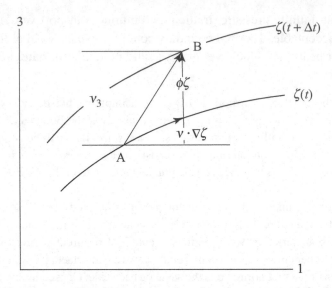

Figure 2.6. At a moving internal interface between saturated freshwater and saltwater regions, for example, the fluid elements on either side must remain in contact, with relative tangential motion in the interface and equal vertical motion on both sides.

Equation (2.27) can be derived more formally by considering the total time derivative or the derivative following the motion of $z = \zeta(x, y, t)$, which moves with the interstitial fluid velocity. From the formula for the total derivative in multivariate calculus,

$$\frac{dz}{dt} = \frac{\partial \zeta}{\partial t} + \frac{\partial \zeta}{\partial x}\frac{dx}{dt} + \frac{\partial \zeta}{\partial y}\frac{dy}{dt},$$

and since $\bar{v}_Z = dz/dt$, and $\bar{v}_H = (dx/dt, dy/dt)$, we recover (2.27).

If the "free" surface is in fact the water table, there may be the additional factor of infiltration from rainfall at the rate $W(t)$, measured in terms of length per unit time, i.e. velocity. This is generally considered as positive even though the rain comes from above, and represents an additional transport to the interface. Thus (2.28) becomes

$$W(t) + w = \phi\dot{\zeta} + \mathbf{u} \cdot \nabla\zeta. \tag{2.29}$$

Some particular cases: with a horizontal surface and no internal flow, $\dot{\zeta} = W(t)/\phi$; in a steady state (or over a time average) in an aquifer recharge region, the vertical transport velocity just below the surface

$$w = -W(t) + \mathbf{u} \cdot \nabla\zeta, \tag{2.30}$$

both terms on the right-hand side being negative since the horizontal flow at the free surface is directed down-slope.

2.5 Mechanical energy balances

2.5.1 Flow tubes and flow resistance

Useful and general hydrological insights can be obtained by considering the overall forcing, flow, resistance and overall energetics in a thin flow tube extending from the water table at a recharge region through a region of possibly inhomogeneous permeability, to its discharge at a lower elevation. The lateral boundaries of the flow tube are defined by streamlines of the flow. Let s be the distance along the axis of the flow tube from the entry point and let $A(s)$ be its cross-sectional area. The permeability can vary along the length of the flow tube, so that $k = k(s)$, in general. Under steady conditions (or in the mean) the incompressibility condition (2.9) ensures that the volume flux through the flow tube $q = u(s)A(s)$ is constant along its length. The Darcy equation (2.24) can be rearranged to express the force balance among the driving pressure gradient, the buoyancy force and the flow resistance,

$$-\nabla p/\rho_0 + b\mathbf{l} = \nu\mathbf{u}/k(s).\tag{2.31}$$

This can be integrated along a flow tube axis from any one designated point, 1, say, to another, 2:

$$-\int_1^2 \nabla(p/\rho_0)\cdot d\mathbf{s} + \int_1^2 b\mathbf{l}\cdot d\mathbf{s} = \nu\int_1^2 k^{-1}\mathbf{u}\cdot d\mathbf{s}.\tag{2.32}$$

The first line integral involving the pressure, integrates to $\rho_0^{-1}[p_1 - p_2]$; the pressure *difference* between the two end points. If, in particular, points 1 and 2 are at an aquifer recharge point and discharge point respectively, this reduces to $g(\zeta_1 - \zeta_2)$, the hydrostatic head difference that drives the flow, as indicated in Figure 2.7. The second integral, which involves the internal buoyancy distribution that may arise from temperature or salinity variations in the flow region, can be written as

$$B = \int b\cos\theta\,ds,$$

where θ is the angle between the flow direction and the upward vertical. This is interpreted as the net buoyancy force along the flow tube axis, which is driving when b and $\cos\theta$ have the same sign (i.e. positive buoyancy in an upward flow or negative buoyancy – excess weight – in a downward flow) or retarding when the

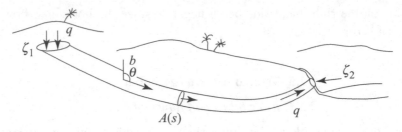

Figure 2.7. A flow tube with cross-section $A(s)$ and constant volume flux discharging into a lake or stream is driven by the hydraulic head of the water table in the infiltration region above the water level at discharge, plus any variations in fluid buoyancy along the tube, and is resisted by the flow tube resistance along the path.

signs are mixed. In the last term, $\mathbf{u} \cdot d\mathbf{s} = (q/A(s)) \, ds$, where q is the volume flux, which is constant along the length of the tube even though the cross-sectional area may vary. Equation (2.32) then becomes

$$\rho_0^{-1}(p_1 - p_2) + B = qR, \qquad (2.33)$$

where the flow tube resistance along the streamline is

$$R = v \int \frac{ds}{k(s)A(s)}. \qquad (2.34)$$

Equation (2.33) therefore expresses the two driving forces, the pressure difference between the two ends of the flow tube and the net internal fluid buoyancy B, which balance the viscous retardation of the fluid in the medium, proportional to the volume flux q. The flow tube resistance has the physical dimensions of $(\text{length} \times \text{time})^{-1}$. Four factors influence its magnitude: the fluid viscosity, the matrix permeability, the path length and the flow tube area. With given hydraulic and thermal driving forces, sections of high permeability provide small flow resistance, even over long path lengths. A "retarding layer" of low permeability may separate aquifers above and below, yet provide only moderate flow resistance if it is relatively thin (a small path length) and if the cross-sectional area of the flow tube is able expand sufficiently, as the fluid seeps across.

A useful quantity in many applications is the flow tube conductance C, which is the reciprocal of the flow tube resistance:

$$C = R^{-1} = \left\{ v \int \frac{ds}{k(s)A(s)} \right\}^{-1}. \qquad (2.35)$$

Equation (2.33) can be rewritten as

$$q = C \left\{ \rho_0^{-1}(p_1 - p_2) + B \right\}, \qquad (2.36)$$

which expresses the volume flux as the flow tube conductance times the sum of the driving forces. If there are a number of flow tubes in parallel with the same driving forces, the flux in each is proportional to the conductance and the total flux is proportional to the sum of all the flow tube conductances. Low-permeability, low-conductance inclusions may have very little effect on the overall volume flux, provided the flow tubes can detour around them, as they may through a gap in an otherwise confining layer. The flow simply avoids regions of the lowest permeability. In any flow domain, the hydrologically important regions are those of high conductance, whose permeability is greatest and connectivity most extensive.

2.5.2 Energy balances

The mechanical energy balance in the interior of the flow can be derived in a similar fashion. The primary energy supply is the inflow of potential energy associated with infiltration of the water at higher reduced pressure (or water table elevation) relative to the discharge, and is given by $\rho_0^{-1} q(p_1 - p_2)$. There may also be internal energy sources or sinks associated with the buoyancy distribution. The internal redistribution of this energy flux can be exhibited by taking the scalar product of the transport velocity $\mathbf{u}$ with Darcy's equation for the force balance in the form (2.31). From elementary calculus, $\mathbf{u} \cdot \nabla p = \nabla \cdot (p\mathbf{u}) - p\nabla \cdot \mathbf{u}$ and $\nabla \cdot \mathbf{u} = 0$ because of incompressibility. This leads to the mechanical energy balance throughout the flow region,

$$-\rho_0^{-1} \nabla \cdot (p\mathbf{u}) + b(\mathbf{l} \cdot \mathbf{u}) = \nu u^2 / k. \tag{2.37}$$

The first term in this equation specifies the spatial rate at which the potential energy is decreasing in the flow direction, and the second expresses the rate at which energy is supplied or extracted from the flow by buoyancy, according to the sign of the buoyancy and the direction of flow. The final term, which is always positive, is the rate ε at which energy per unit mass of interstitial fluid is dissipated by the viscous flow through the interstices:

$$\varepsilon = \nu \frac{u^2}{k}, \tag{2.38}$$

where ν is the kinematic viscosity.

The integrated version of the mechanical energy balance (2.37) over a thin flow tube of cross-sectional area $A(s)$ is also of interest. Since, again, the volume flux $q = u(s)A(s)$ is constant along the flow tube and with use of (2.24), the integrated equation becomes

$$\rho_0^{-1} q(p_1 - p_2) + q \int b \cos \theta \, ds = q^2 R \tag{2.39}$$

in terms of the flow tube resistance R in (2.34). In this integral energy balance, the rate of input of potential energy that drives the flow plus the rate of energy supply from buoyancy is balanced by the internal rate of energy dissipation $q^2 R$. Note that the internal pressure distribution has disappeared from this overall balance since it provides for the flux of energy from one point to another throughout the interior, but is not an internal energy source or sink. The overall energy balance is the integral of (2.37) over the whole flow domain (i.e. over all the flow tubes):

$$-\rho_0^{-1} \int p\mathbf{u} \cdot ds + \int bw dV = \int \varepsilon dV. \qquad (2.40)$$

The three terms represent, in order, the net potential energy input associated with infiltration into the region ($\mathbf{u} \cdot ds < 0$) at higher reduced pressures and discharge at lower ones, the energy supplied by the upward motion of more buoyant fluid in the interior, or by downward motion of denser fluid, or lost by the reverse, and finally, the viscous dissipation throughout the entire domain.

In these derivations, the matrix and the fluid have been regarded as incompressible. Transient compressibility effects can be important following seismic events as discussed in Section 3.6, but usually, when compared with the gravitational potential energy, the strain energy of a hydrological system (the energy of elastic compression of the fluid) is a trivial part of the overall energy balance. The strain energy per unit mass of water is defined as

$$E_S = -p\frac{dV}{V} = p\frac{d\rho}{\rho},$$

where dV and $d\rho$ are the changes in specific volume and density produced by the pressure p. Changes in pressure and density are related by $dp = c^2 d\rho$, where c is the speed of sound, which, for the most compressible constituent, water, is about 1400 m/s. In a permeable formation extending to a depth h, the fluid pressure is approximately $\rho g z$, where z is measured down from the water table, and the volumetric strain in the fluid at depth z is $d\rho/\rho = dp/\rho c^2 = gz/c^2$. Consequently the mean strain energy of the water (occupying the volume fraction ϕ of the water column) is

$$E_S \sim \tfrac{1}{2}\phi\rho g h(gh/c^2)$$

while the mean gravitational potential energy $E_P = \tfrac{1}{2}\rho g h$ (the numerical factor arising because the mean pressure is half the base value). The ratio of mean strain energy to mean gravitational potential energy is equal to $\phi g h/c^2$. Even if the vertical extent h of the aquifer is as large as 1 km, this ratio is less than 1%. The fluid kinetic energy is far smaller still.

2.6 Two theorems

This section concerns the application of two theorems that were established by Helmholtz (see Lamb, 1932), concerning the flow of viscous fluid of constant density at very low Reynolds number. For a modern discussion, see Batchelor (1967). In principle, Helmholtz's general results for flow in arbitrary connected domains do cover interstitial flow in permeable media with arbitrary distributions of (positive) permeability, but direct proofs for Darcy flow are very much simpler. Despite their antiquity, the theorems seem to have been largely overlooked in this area, though they do offer the potential for many conceptual insights and practical applications.

2.6.1 The uniqueness theorem

The uniqueness theorem established below asserts that there cannot be more than one Darcy flow solution for *constant density flow* with given boundary conditions – the solution is unique. The immediate utility of this theorem is that, if one finds a solution to such a problem, there are no others. The solution cannot develop a bifurcation or an instability. The theorem is true no matter what the internal permeability distribution may be. Note, however, that this uniqueness applies to fluids of constant density; it *does not extend to situations in which buoyancy forces are involved*. In fact, we know by examples that it is not true in these situations. One simple such case involves a saturated permeable region with horizontal isopycnals (lines of constant density). One solution is a state of rest – the velocity is everywhere zero. When fresh water lies over more saline water, the density decreases with height and the state of rest is stable – if it is disturbed slightly, it will return to its initial state of rest. However, when the salinity and density *increase* with height, the state of rest is unstable; tiny perturbations amplify and the system moves to a new time-dependent or steady-state solution.

To prove the uniqueness for constant density Darcy flow, consider the flow through a medium in which the permeability $k(\mathbf{x})$ is an arbitrary (but positive) function of position. In the absence of buoyancy forces,

$$\mathbf{u} = -(k(\mathbf{x})/\mu)\nabla p, \tag{2.41}$$

where p is the reduced pressure. The incompressibility condition is $\nabla \cdot \mathbf{u} = 0$. Let us suppose that the theorem is false, that two different patterns of flow and pressure $(\mathbf{u}, p)$ and $(\mathbf{u}', p')$ are possible in a given region V, both satisfying (2.41) and the incompressibility condition, with assigned distributions of either pressure or normal component of transport velocity on the boundaries S;

$$p = p' \quad \text{or} \quad \mathbf{u} \cdot d\mathbf{S} = \mathbf{u}' \cdot d\mathbf{S} \text{ on the boundary surface } S. \tag{2.42}$$

We prove that the two solutions are, in fact, the same. Consider the following integral throughout the volume

$$\int k(\mathbf{x})[\nabla p - \nabla p']^2 dV$$

$$= \int k(\mathbf{x})[\nabla p - \nabla p'] \cdot [\nabla p - \nabla p'] dV,$$

$$= -\mu \int (\mathbf{u} - \mathbf{u}') \cdot \nabla(p - p') dV, \text{ from (2.6.1)},$$

$$= -\mu \int \nabla \cdot [(\mathbf{u} - \mathbf{u}')(p - p')] dV, \text{ from incompressibility},$$

$$= -\mu \int (p - p')(\mathbf{u} - \mathbf{u}') \cdot d\mathbf{S}, \text{ from the divergence theorem}$$

$$= 0, \tag{2.43}$$

the last step expressing the identity of boundary conditions (2.42). The first integral (2.43) therefore vanishes and since $k(\mathbf{x})$ is everywhere positive, it follows that the integrand must be zero and consequently $\nabla p = \nabla p'$ everywhere. The two solutions are identical and the theorem is true. Solutions are unique.

2.6.2 *The minimum dissipation theorem*

The minimum dissipation theorem is not only conceptually important but also useful in some practical situations by providing a means of inferring, without detailed calculation, the general nature of flow patterns in perhaps complex geological situations. The theorem statement is as follows. In a region occupied by a permeable medium in which buoyancy effects are negligible, if the transport velocity distribution over the boundary of the region is prescribed, then the actual internal flow has a total rate of dissipation of energy that is less than any other conceivable, kinematically possible flow in the same region with the same boundary conditions. (A kinematically possible flow is any velocity field one might imagine that satisfies the incompressibility condition, but not necessarily the Darcy equation. The actual flow, of course, satisfies both.) For example, in a near-surface groundwater flow, rainwater infiltrates downward to the water table at a rate that can be regarded as given, moves a possibly significant distance as groundwater through a matrix with a complex distribution of permeability and ultimately discharges into streams or lakes. Whatever the internal permeability structure may be, this remarkable theorem asserts that the patterns of groundwater flow speed and direction are those which minimize the overall dissipation rate.

The proof is as follows. Let $\mathbf{u}(\mathbf{x})$, $p(\mathbf{x})$ represent the true solution with $\mathbf{u}$ prescribed on the boundary S. Consider a kinematically possible alternative flow that one might dream up, $\mathbf{u} + \mathbf{u}'$, $p + p'$, with the same velocities at the boundary, so that

$$\mathbf{u}' = 0 \text{ on } S. \qquad (2.44)$$

This alternative flow also satisfies the incompressibility condition so that $\nabla \cdot \mathbf{u}' = 0$, but it may not satisfy the Darcy equation. From (2.36), the total rate of energy dissipation in the alternative flow is

$$\int \varepsilon dV = \mu \int \{(\mathbf{u} + \mathbf{u}')^2/k\}dV,$$
$$= \mu \int \{(\mathbf{u}^2 + \mathbf{u}'^2)/k\}dV + 2\mu \int (\mathbf{u} \cdot \mathbf{u}'/k)dV.$$

Since the true solution does satisfy the Darcy equation $\mu \mathbf{u}/k = -\nabla p$, the last term above becomes

$$2 \int (\nabla p) \cdot \mathbf{u}' dV = 2 \int \nabla \cdot (p\mathbf{u}')dV \quad \text{since } \nabla \cdot \mathbf{u}' = 0,$$
$$= 2 \int p\mathbf{u}' \cdot dS \quad \text{from the divergence theorem,}$$
$$= 0, \quad \text{from (2.44).}$$

Accordingly, the total dissipation rate in the alternative flow is $\mu \int \{(\mathbf{u}^2 + \mathbf{u}'^2)/k\}dV$, which is greater than that of the true flow, namely $\mu \int (\mathbf{u}^2/k)dV$. This establishes the theorem.

It is a very important and far-reaching result. Examination of the derivation above confirms that it remains true if the permeability $k(\mathbf{x})$ in the region is not uniform provided only that the velocity distribution across the boundaries is specified. Regions or lenses of low permeability are generally sites of proportionately low flow velocities, and because of the quadratic dependence of the dissipation on $\mathbf{u}$, they contribute relatively little to the overall dissipation. The flow occurs preferentially in the high-permeability, low-resistance regions, attesting to the accuracy of the old adage that "flow follows the path of least resistance." It is a robust theorem, finding a number of applications in later sections of this book.

2.7 The thermal energy balance

The distribution of temperature in the fabric is constrained by an equation describing the heat balance in each averaging volume – an expression of the first law of thermodynamics. Let C represent the specific heat at constant pressure; we will use subscripts "M" and "F" to refer to properties of the saturated matrix as a whole and to those of the fluid, respectively. The rate of change in time of the heat content in any element of unit volume is $(\rho C_M)\partial T/\partial t$, and this may come about as a result of heat addition (or subtraction) brought about by convective heat transport in moving interstitial fluids, by molecular conduction through the matrix, and possibly by heat generation (or absorption) in chemical reactions.

The convective heat flux across unit area of the fabric is given by the product of the heat content per unit volume of the interstitial fluid $(\rho C)_F T$, times the mean interstitial fluid velocity $\bar{\mathbf{v}}$ with which it moves, times the fraction ϕ of the area occupied by the fluid, that is, by $(\rho C)_F \mathbf{u} T$. The rate of accumulation of heat in the volume element associated with this is the *net* flux inwards, or minus the net flux outwards across the surfaces of the element of volume,

$$-(\rho C_F) \int (T\mathbf{u}) \cdot \mathbf{n} dS = -(\rho C_F) \int \nabla \cdot (T\mathbf{u}) dV,$$

from the divergence theorem. The convective heat input per unit volume is consequently

$$-(\rho C)_F \nabla \cdot (\mathbf{u} T) = -(\rho C)_F \mathbf{u} \cdot \nabla T, \tag{2.45}$$

in virtue of the incompressibility condition (2.8).

Heat is also transferred by conduction through the saturated matrix. The conductive heat flux down the gradient is proportional to the magnitude of that gradient – this is Fourier's law of heat conduction – and can be expressed as $-\kappa_M \nabla T$, where κ_M is the thermal conductivity of the saturated medium. The rate of accumulation of heat per unit volume is then minus the divergence of this, or

$$\nabla \cdot (\kappa_M \nabla T) = \kappa_M \nabla^2 T, \tag{2.46}$$

when the medium is thermally homogeneous. Finally, if chemical reactions are taking place, heat may be added at the rate $\hat{Q}$ per unit volume. Combining these expressions, we have the thermal energy balance

$$(\rho C)_M \frac{\partial T}{\partial t} = -(\rho C)_F \mathbf{u} \cdot \nabla T + \kappa_M \nabla^2 T + \hat{Q}. \tag{2.47}$$

This can be rearranged slightly and divided throughout by $(\rho C)_F$ to give

$$M \frac{\partial T}{\partial t} + \mathbf{u} \cdot \nabla T = \kappa \nabla^2 T + Q, \tag{2.48}$$

where $M = (\rho C)_M/(\rho C)_F$, the heat source $Q = \hat{Q}/(\rho C)_F$, and the thermal diffusivity κ, with physical dimensions $(\text{length})^2 \times (\text{time})^{-1}$, is the matrix thermal conductivity divided by $(\rho C)_F$. The relative density of most crustal rocks is about 2.6, while the specific heats are between 0.19 and 0.21, so that M is generally 0.5 ± 0.1. Small glass beads are sometimes used as the porous medium in convection experiments, and for them, $M \sim 0.4$. The form of the advection term in the heat conservation equation above indicates that in a permeable, water-saturated medium, the effective advection velocity for heat is $\mathbf{u}/M$, somewhat larger than the transport velocity but smaller than the interstitial fluid velocity $\mathbf{u}/\phi$. This is because heat is diffused (or "leaked") from the fluid pathways into the matrix solids, while the interstitial fluids and inert solutes remain in the interstices.

While the advection of heat by the mean flow is always important in thermally driven flows, the dispersion of heat about the mean flow streamlines by the random excursions of fluid among the grains seems less so, except possibly in more permeable media such as insulating material (Kvernvold and Tyvand, 1980). The thermal diffusivity for saturated rock, κ is of the order $10^{-7}\,\mathrm{m^2 s^{-1}}$ (MKS units), about the same as that for water. Heat is conducted through the entire fabric, not just along the fluid pathways, which occupy only the fraction ϕ of the area across which thermal dispersion occurs. Molecular heat conduction through the fabric dominates the random heat dispersion on the pore scale l about the mean streamlines when

$$\kappa \gg \phi\bar{v}l, \tag{2.49}$$

a condition that is usually satisfied in geological contexts. For example, even in a sandy aquifer with porosity $\phi \approx 0.3$, pore diameters of order $10^{-3}\,\mathrm{m}$ and a mean interstitial fluid velocity of 1 km/yr $\sim 3 \times 10^{-5}\,\mathrm{m/s}$ (which for most aquifers is a high value) the left-hand side of (2.49) is about 10 times larger than the right.

2.8 Dissolved species balance

The patterns of flow through sediments are of interest to the environmental engineer in questions of the dispersal of reactive chemical wastes in surface aquifers and to the geochemist largely because of the physical and chemical changes in the fabric that are being produced or were produced in the distant past. Both are influenced by cementation in which solid material is deposited from solution at the pore boundaries, by dissolution when it is removed and carried away, or by various chemical reactions between the solid constituents of the fabric and the solutes of the interstitial fluid.

Equations for the balances of chemical constituents in the interstitial fluid will be developed in the same way as in the preceding section. Since our interest is in *spatial* distributions of chemical reactions and reaction products, the interstitial fluid concentrations, $c(\mathbf{x}, t)$, are defined as mass of dissolved solute per unit *volume* of solution, with dimensions $[ML^{-3}]$, rather than per unit mass, which is dimensionless. The solute concentration balance, like the mass balance (2.11) and the heat balance (2.48), has the same form regardless of the units in which it is expressed, whether moles or mass per unit volume, provided they are consistent throughout.

The spatial distributions of fabric alteration involve processes on a wide range of scales. (i) On the largest scales, from that of the formation itself to the matrix averaging scale (Section 2.2), solutes are transported advectively by the mean interstitial fluid flow $\mathbf{v}(\mathbf{x}, t)$ through the pores and fractures in the medium. (ii) Within the matrix averaging volumes, solute is dispersed in spatially random fluid pathways and possibly by small-scale layering, while finally, (iii) on the scale of the pores themselves, molecular diffusion to and from the pore walls allows sorption or interfacial chemical reaction between the solid matrix and the pore fluids, as described by Compton and Unwin (1990). Only the first of these is resolved explicitly in the solute balance equations, with dispersion within the matrix averaging volumes being expressed in terms of a dispersivity and the chemistry being specified in terms of kinetic reaction rates and solution concentrations.

Thus, the mass of a particular solute in the interstitial fluid per unit volume of the matrix is $\phi c(\mathbf{x}, t)$, where ϕ is the matrix porosity, and its rate of change is the net result of (i) the advective flux convergence $-\nabla \cdot \{\phi \mathbf{v}(\mathbf{x}, t)c(\mathbf{x}, t)\} = -\nabla \cdot (\mathbf{u}c)$, where $\mathbf{u} = \phi \mathbf{v}$ is the transport velocity, and (ii) small-scale dispersive flux $\phi D \nabla c$, where D is the macroscopic dispersion coefficient (Section 2.10, below). The dispersive flux *divergence* is then $\phi D \nabla^2 c$. Finally, (iii) for the present, the sub-pore and molecular-scale processes will be expressed as an overall source term ϕQ_C, where Q_C represents the mass of solute added per unit volume of the *fluid* per unit time, which depends on the chemistry, the surface properties of the internal fluid/solid interfaces and their microscale geometry. Accordingly, the species balance for each dissolved constituent is represented by

$$\phi \frac{\partial c}{\partial t} = -\nabla \cdot (\mathbf{u}c) + \phi D \nabla^2 c + \phi Q_C. \tag{2.50}$$

If the rate of generation of fluid in the reaction is zero or negligible compared with the external fluid flux, then $\nabla \cdot \mathbf{u} = 0$ and the equation can be written in terms of the mean interstitial fluid velocity as

$$\frac{dc}{dt} = \frac{\partial c}{\partial t} + \bar{\mathbf{v}} \cdot \nabla c = D \nabla^2 c + Q_C, \tag{2.51}$$

The advection velocity for non-reacting dissolved species is the mean interstitial velocity $\bar{\mathbf{v}} = \mathbf{u}/\phi$, and (2.51) states that for fluid elements moving with this velocity, the rate of change following the motion of solute concentration dc/dt is the result of both dispersion and chemical reaction. Sorbtion and chemical reaction with the matrix generally reduce the effective advection velocity, as described in Chapter 5. The difference between the effective advection velocities for heat and dissolved salts is brought about by the fact that advected heat changes the temperature of the whole matrix whereas non-reacting advected salts remain in the fluid phase. This difference has profound consequences in the transport of dissolved contaminants and in thermo-haline instabilities as described in Chapters 4 and 5.

The salinity S of the interstitial fluid is usually defined as the total dissolved mass per unit *mass* of solution; when a number of dissolved species is present $S = \rho^{-1}\Sigma c$, summed over the concentrations c of the separate constituents. A conservation equation for S follows from (2.51); since the equation is linear in c, we have

$$\frac{\partial S}{\partial t} + \bar{\mathbf{v}} \cdot \nabla S = D\nabla^2 S + Q_S \tag{2.52}$$

where the combined source term $Q_S = \rho^{-1}\Sigma Q_C$ may be nonlinearly dependent upon the individual concentrations.

2.8.1 Rate-limiting steps and the solute source term

The generic expression Q_C in (2.50) and (2.51) may represent the addition of solute to the interstitial fluid by one or more of a variety of physical, chemical or biological processes. Dissolution of the solid phase adds solute to the interstitial fluid and $Q_C > 0$, while precipitation depletes it, $Q_C < 0$. Oxidation or reduction reactions or organic decomposition may provide both sources and sinks. Most of the reactions of interest in this book occur at the solid/liquid interfaces at the surfaces of pores or fractures, on scales much smaller than the averaging volume that is implied in the continuum representation of equations like (2.52). This suggests that, among many other factors, Q_C is proportional to the average area of reacting surface per unit volume of fabric, or approximately, to the volume fraction of the reacting mineral. Nevertheless, the Darcy-type averaging process distributes the sources throughout the averaging volume and Q_C becomes a field variable, determined in detail by interfacial surface morphology, the chemical kinetics (Lerman, 1979, Drever, 1982, Helgeson, Murphy and Aagaard, 1984, and others) together with other molecular processes.

Even in a process so apparently simple as dissolution, the accepted scheme, as described by Compton and Unwin (1990), involves a sequence of steps including (*a*) advective and diffusive transport of the reactant from its source to the reaction location, (*b*) adsorption of the reactant onto the surface, (*c*) diffusion of the reactant on the surface to a reactive site, (*d*) reaction, (*e*) diffusion of products away from the immediate reaction site, (*f*) product desorption, and (*g*) diffusive and advective transport of the reaction products away from the reaction site. In the context of reactions between a solid matrix and interstitial fluids, the processes (*a*) and (*g*) are site-specific and specified explicitly by the advective and diffusive terms in equation (2.51). The ionic concentration of interstitial fluid is usually several orders of magnitude smaller than in the solid minerals with which reaction occurs, and although it may take decades or centuries for interstitial aqueous solutions to pass through a geological formation, the time required to supply mineralogically significant amounts of reactant is *even greater* by a further several orders of magnitude. The time required to progress through the sequence of internal steps, (*b*) to (*f*), is determined essentially by the time scale over which the *slowest* or *rate-limiting* step in the ladder occurs, and this is frequently much smaller than the reactant transport time. The reactions occur more rapidly than does the transport of reactant to the reaction site. In this circumstance, the reaction region is concentrated into a relatively narrow zone or front that propagates through the region in the direction of the mean interstitial flow and separates the region of unaltered rock ahead from the region behind in which the reaction has completed. This important process is described in quantitative detail in Section 5.4.

The speed at which the reaction front propagates is proportional to the supply of reactant, but the internal structure of the front involves the sequence of internal steps, (*b*) to (*f*), listed above. Within these, the rate-limiting step seems frequently to be determined by the reaction kinetics (a function of the activities involved) at activated sites on the surface of the solid mineral, so that the active interstitial surface area per unit volume of the fabric is also an important factor. In the present context, molecular diffusive exchange between the pore or fracture volume and the adjacent surface is rapid, taking only seconds or minutes in a 10 micron pore, and only seldom may be limiting. The reaction progress can represented in terms of an overall rate constant γ, which lumps together these factors and the temperature dependence (and consequently varies widely), multiplied by a function of the relative saturation c/c_E, where c_E is the equilibrium concentration:

$$Q_C = \gamma c_E f(c/c_E) \tag{2.53}$$

where γ has dimensions $(\text{time})^{-1}$ and the function f is dimensionless. The form of the function f can be determined experimentally; it may be dependent on pH or the concentration of other ions present. Over restricted ranges of the argument, it can sometimes be represented as a power-law relationship, and the power involved is called the order of the reaction. For example, the experiments of Plummer and Wigley (1976) show that calcite in water at 25 °C and 1 atmosphere partial pressure of CO_2 dissolves as a second-order reaction as the pH increases from 3.9 when $c = 0$, to 5.9 as the calcite dissolves.

There are a few simple constraints on the form of the function f that determine the reaction progress. When a solid is dissolving, the rate of dissolution, proportional to f, is expected to be a maximum when the surrounding solution is extremely dilute, so that we can take

$$f(0) = 1, \qquad Q_C = \gamma c_E \quad \text{when} \quad c/c_E \to 0. \qquad (2.54)$$

Note that this condition in fact defines γ as the actual dissolution rate in a very dilute solution, divided by the equilibrium fluid concentration, both of which are measured quantities. When the solution is saturated at a given temperature and pressure, a state of equilibrium exists between solid and saturated solution and the net source vanishes:

$$f(1) = 0; \qquad Q_C = 0 \quad \text{when} \quad c = c_E. \qquad (2.55)$$

When a solution is supersaturated $c > c_E$, the function f is negative, and solute precipitates or crystallizes from the solution. A simple form of the distribution function f that satisfies these conditions is

$$f(c/c_E) = 1 - (c/c_E)^n, \qquad n > 0, \qquad (2.56)$$

though it has little other chemical justification. Near equilibrium, i.e. when $c = c_E - \Delta c$, where $\Delta c/c_E$ is small (2.56) reduces to

$$f(c/c_E) \approx n(\Delta c/c_E) \quad \text{and} \quad Q_C \approx n\gamma(\Delta c);$$

the rate of dissolution is proportional to the index n, the reaction rate and the under-saturation Δc. Since the overall rate "constant" γ is a lumped parameter in geological applications, its value can vary widely even with the same mineralogy. Lerman (1979) quotes values of γ from laboratory measurements on silicate minerals from 2 to 20 yr^{-1} or $(0.6-6) \times 10^{-7}$ s^{-1}, whereas those inferred from deep ocean cores in siliceous sediments are several orders of magnitude smaller!

In the species concentration balance, (2.51), the source or chemical reaction term is of general order γc since the distribution term is of order unity, $f(c/c_E) \sim 1$. The advection term is of order $\bar{v}c/l$ where $\bar{v}$ is the mean interstitial fluid velocity and l, the scale of variation of the fluid concentration in the flow direction. The ratio

$$Da = \frac{\gamma l}{\bar{v}}, \qquad (2.57)$$

known in chemical engineering as the Damköhler number, is the ratio of the time $(l/\bar{v})$ taken for the interstitial fluid to move a distance l through the matrix, to the time scale for chemical reaction γ^{-1}. In a reaction front, these two quantities are equal so that $Da \sim 1$ and $l \sim \bar{v}/\gamma$ characterizes the thickness of the front.

During the 1990s it became increasingly apparent that unicellular microorganisms, metal-reducing bacteria are common in groundwater and sediments, and are extremely efficient catalysts in a variety of geochemical reactions. Lovley *et al.* (1989) showed that even under anoxic conditions, these organisms catalysed the reduction of Fe^{3+} ions and the oxidation of aromatic hydrocarbons in groundwater. Geologic Fe^{3+} oxides showed a large range of reducibility, approaching 100% in some materials (Zachara and Rodin, 1996), and Russell *et al.* (2003) claim that hydrolysis rate enhancements of order 10^6 can be achieved. A brief but useful review has been given by Newman (2003). The mechanisms of electron exchange are not yet entirely clear, and the range of materials in which they occur is somewhat uncertain. However, when reactions can be catalyzed effectively in this way, the kinetic reaction rates may be so rapid that the reaction progress is limited by other steps in the sequence. The principal consequence of this catalysis may well be a widespread reduction in the thickness of reaction fronts with their propagation rate unchanged, since the front propagation speed is dependent on the supply of reactant, not the speed at which the reaction occurs.

Somewhat paradoxically, one geochemical reaction scenario in which the reactions are clearly free of flow control involves the hot springs in places like Yellowstone in Wyoming, Rotorua in New Zealand and Pamukkale in Anatolia. Here, warm mineral-laden aqueous solutions bubble to the surface, depositing calcite in pools and terraces at an easily observable rate. The species balance equation in the spring waters is precisely identical to the balance (2.51) for interstitial fluids in the porosity limit $\phi \to 1$, i.e. where the solid matrix disappears and the transport velocity and the mean velocity of fluid elements become identical. Interstitial fluid velocities in pores or fractures may be a few m/yr while the fluid velocities $\mathbf{u} = \overline{\mathbf{v}}$ in the springs can be seen to be characteristically a few cm/s, greater by a factor of about 3×10^9. The length scales of the reaction zones in the two cases may generally be comparable, so that the time scales involved in reactant transport in the hot springs are much smaller that in aquifer interstices and the rates are much larger, by factors of order 10^9–10^{10}. In spite of the generally warmer temperature in the hot springs, the rate-limiting step in the reaction sequence there is very likely to lie in the kinetics, not the flow.

2.8.2 First-order reactions

Near equilibrium, one might expect that the rate of addition of solute to the interstitial fluid by dissolution is linearly proportional to the local concentration difference from the equilibrium value:

$$\mathcal{Q}_C = \gamma(c_E - c), \tag{2.58}$$

which also follows from (2.56) with $n = 1$. Reactions in which the rate of production or disappearance of a dissolved species is linearly proportional to the difference between the local concentration and the equilibrium value are described as first order.

Dissolution and precipitation may occur simultaneously when polymorphic forms of a solid phase have different solubilities. In the case of silica, for example, the solubility of amorphous silica (opal) is considerably greater than that of quartz, those of other polymorphs being intermediate (see, for example, Lerman 1979, p. 387). The silica concentration in the interstitial fluids is then described by

$$\frac{\partial c}{\partial t} + \phi^{-1}\mathbf{u} \cdot \nabla c = D\nabla^2 c + \sum_n \gamma_n (c_{nE} - c),$$

where γ_n is the rate constant and c_{nE} the saturation concentration of the nth polymorph. Alternatively,

$$\frac{\partial c}{\partial t} + \phi^{-1}\mathbf{u} \cdot \nabla c = D\nabla^2 c + \gamma(C - c), \tag{2.59}$$

where

$$\gamma = \sum_n \gamma_n, \qquad C = \gamma^{-1}\left(\sum_n \gamma_n c_{nE}\right). \tag{2.60}$$

In a steady state with no flow, the interstitial fluid concentration stabilizes at $c = C$, which is the arithmetical average of the saturation concentrations of the individual polymorphs, weighted by their respective rate constants. Those for which $c_{nE} > C$ are dissolving while polymorphs for which $c_{nE} < C$ are being precipitated.

2.9 Equations of state

The density of the interstitial fluid, usually an aqueous solution, depends on the temperature T, pressure p, and the concentrations c_n (mass per unit volume) of dissolved solids. The salinity S (dissolved mass per unit mass) is defined as the sum of the solute concentrations,

$$S = \rho^{-1} \sum_n c_n, \tag{2.61}$$

where ρ is the solution density. The relationship among these quantities,

$$\rho = \rho(p, T, S), \tag{2.62}$$

is the equation of state. It is generally nonlinear and, for some fluids, it has not been measured with great precision. The behavior of fresh or low-salinity water in the range $0-10\,°C$ is particularly anomalous. As the temperature decreases from about $10\,°C$, water contracts and increases in density until at $4\,°C$ a density maximum is attained. Below this, the water expands and its density decreases as the temperature falls towards 0 °C. Ice forms first at the water surface.

In geological flow situations, variations in interstitial fluid density are most significant in their influence on the fluid buoyancy – horizontal variations in buoyancy necessarily lead to fluid motion, as will be seen in the next chapter. When the fluid temperature and pressure are sufficiently far removed from the critical and freezing points, and the temperature and salinity ranges in the field of flow are sufficiently small, the equation of state can be represented with reasonable accuracy as a linear function of temperature and/or salinity, such as

$$\rho = \rho_0(1 - \alpha T + \beta S), \tag{2.63}$$

where $\rho = \rho_0(p)$ is the reference density at the ambient pressure and

$$\alpha = -\rho_0^{-1}\partial\rho/\partial T, \qquad \beta = \rho_0^{-1}\partial\rho/\partial S \tag{2.64}$$

are the coefficient of volumetric expansion and density coefficient for salinity, respectively. For saline water in the range 10–20 °C, $\alpha \approx 1.5 \times 10^{-4}(°C)^{-1}$ and $\beta \approx 0.78$ when the salinity is expressed as the dissolved mass of salt per unit mass of solution (Weast, 1972).

However, when significant variations in interstitial density are produced by dissolution of a major fabric component or by ex-solution, the concentration may be close to saturation at the local ambient temperature: $c = c_E(T)$. The equation of state can then be expressed as

$$\rho = \rho_0(1 - \alpha_S T),$$

where

$$\alpha_S = -\frac{1}{\rho_0}\frac{\partial\{\rho(T, c_E(T))\}}{\partial T} \tag{2.65}$$

can be described as the thermal expansion coefficient of the continuously saturated solution. Its magnitude and *sign* depend on the chemical nature of the solute. From (2.65)

$$\alpha_S = -\frac{1}{\rho_0}\left\{\frac{\partial\rho}{\partial T} + \frac{\partial\rho}{\partial c_E}\frac{\partial c_E}{\partial T}\right\},$$

and if the saturated solute (rather than any other that may be present) dominates the variations in density,

$$\frac{\partial\rho}{\partial c_E} = -\frac{1}{\rho_0}\frac{\partial\rho}{\partial S} = \beta.$$

Consequently, the thermal expansion coefficient for a continuously saturated solution is

$$\alpha_S = \alpha - \frac{\beta}{\rho_0} \frac{\partial c_E(T)}{\partial T}. \tag{2.66}$$

Representative values of α, β and $\partial c_E / \partial T$ can be found in the *International Critical Tables* (Washburn, 1929) or the *Handbook of Chemistry and Physics* (Weast, 1972). Solutes that saturate at very low concentrations have $\alpha_S \approx \alpha$, the ordinary coefficient of expansion for water. For most solutions, $\partial c_E / \partial T > 0$, though $Ca_2SO_4 \cdot 2H_2O$ (natural gypsum) is an exception with c_E decreasing from 2.4×10^{-2} g/cm^3 at 20 °C to about 2.2×10^{-3} at 80 °C. Those whose saturation concentration is large and increases rapidly with temperature may have $\alpha_S < 0$, as the temperature increases, so does the density, since the volumetric expansion is more than offset by the increased mass of solute in the concentrated solution.

The fluid viscosity μ is also a function of temperature and, to a lesser extent, of pressure and salinity, and a detailed collation is given by Kestin, Khalifa and Correia (1981). For example, the viscosity of water decreases by a factor of about three from 0 °C (1.79 cp) to 40 °C (0.65 cp). However, the viscosity (appearing in Darcy's law) always occurs in combination with the permeability, whose variability in natural formations is usually much greater than the variations in fluid viscosity.

2.10 Dispersion

The spatially random fluid motion in a porous medium with variable or random permeability is much more tightly constrained than is turbulent motion in a river or lake. In an Eulerian frame of reference, fixed with respect to the matrix, the velocity field is essentially steady in time, although it may vary on seasonal or climatic time scales. The constraint of uniqueness insists that, in the absence of buoyancy effects, there is only one flow solution that cannot evolve in time to another. The minimum dissipation theorem severely restricts the meandering of the flow in a randomly permeable medium. Only when it offers lower overall dissipation can a streamline deflect from the shortest path, in order to detour through a region of higher permeability or to deflect around one of lower permeability. There are no eddies in groundwater flow patterns.

Nevertheless, it is intuitively evident that as marked, interstitial fluid elements move through an aquifer with randomly varying permeability or an extensive fracture network, they will tend to disperse about their mean pathways. The velocity field is spatially random but approximately steady in time. In a sufficiently large and generally homogeneous domain, variations in local permeability produce distributions that tend to become Gaussian (Fickian) with the root mean square spread about the mean streamline that is asymptotically proportional to $(Dt)^{1/2}$, where D is the appropriate diffusivity and t is the elapsed time. This asymptotic state can be attained with an effective diffusivity proportional to the product of the length scale of the *variations* in permeability and the consequent fractional variations in interstitial fluid velocity, provided the flow domain size is much larger than the characteristic scale of the permeability variations.

The structural geology has a profound influence on dispersal. In a fracture–matrix medium, fluid in the fractures moves and disperses more rapidly than fluid in the blocks, so that a pattern of double dispersion emerges. When the dispersal is a consequence of a more-or-less random distribution of more permeable lenses that concentrate and redistribute the flow, the effective pathway intersection scale may be comparable with the flow domain size, and the $(Dt)^{1/2}$ asymptotic state is not attained at all. In these circumstances, the dispersion is scale dependent, with a root mean square particle displacement from the mean that increases in time at rates between $t^{1/2}$ and t. An understanding of dispersion when the scale of inhomogeneity is not very small compared with the flow domain size, requires much more detailed specification of the large-scale permeability structure than is needed above. Important contributions to understanding the relationships between dispersion and the statistics of large-scale anisotropy have been made by Dagan (1982, 1984), who called the phenomenon "mega-dispersion."

2.10.1 Kinematics of dispersion

Consider the dispersal of individual elements of fluid or contaminant as they move through the spatially random permeability variations in a sandy aquifer or the intersecting conduits in an extensively fractured medium. Two separate questions are involved. The first concerns only the flow geometry or kinematics: what particular characteristics of the spatially random interstitial velocity field determine this dispersal? This is considered below. The second question involves the specific dynamical interactions between the flow and the random distribution of permeability as measured by Hess, Wolf and Celia (1992), and others. What are the velocity variations produced by the largely horizontal mean flow moving through this distribution? This is discussed in Section 3.3, below.

The basic Lagrangian analysis of fluid dispersal follows that of G. I. Taylor (1921) on "diffusion by continuous movements" in turbulent flow, where the locations x of individual fluid elements in a small averaging volume were traced in terms of their initial positions a and the elapsed time, t. The same approach can be applied to the dispersion of an injected pulse of marked or contaminated fluid by the spatially random but time independent velocity field in a natural geological

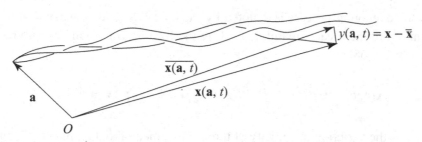

Figure 2.8. In the Lagrangian specification of fluid motion, each individual fluid element is tagged by its initial position **a** relative to a fixed origin and its motion expressed as the time derivative of its subsequent displacement $\mathbf{x} = \mathbf{x}(\mathbf{a}, t)$.

structure. As a marked fluid element moves through the medium, its trajectory is a function of its initial position and the elapsed time, $\mathbf{x} = \mathbf{x}(\mathbf{a}, t)$, as illustrated in Figure 2.8. Note that the initial instant, $t = 0$, the fluid element is at **a**, so that $\mathbf{x}(\mathbf{a}, 0) = \mathbf{a}$. The interstitial velocity *of the fluid element* is the time derivative of its position:

$$\frac{d}{dt}\{\mathbf{x}(\mathbf{a}, t)\} = \mathbf{v}(\mathbf{a}, t), \tag{2.67}$$

and this is also a function of initial position and time as we follow it along. Equation (2.67) can be expressed equivalently in a useful integral form:

$$\mathbf{x}(\mathbf{a}, t) - \mathbf{a} = \int_0^t \mathbf{v}(\mathbf{a}, t')dt', \tag{2.68}$$

where t' in the integral is the elapsed time. This gives the displacement of the fluid element from its initial position as an integral of its velocity along its possibly convoluted trajectory. This and other fluid elements in the same averaging volume (which on the macroscopic resolution scale are at the same point) have trajectories that diverge in time as some move one way around the local inhomogeneities and others, another. From a given averaging volume, then, there is an ensemble of trajectories, which is equivalent to the result of marking the entire fluid in the volume and following the subsequent history of the cloud.

The mean streamline through the point **a** is defined by the mean displacement of the centroid of the marked fluid from its initial position over a fixed time interval t, and averaged over the cloud.

$$\overline{\mathbf{x}(\mathbf{a}, t)} - \mathbf{a} = \overline{\int_0^t \mathbf{v}(\mathbf{a}, t')dt'}$$

$$= \int_0^t \overline{\mathbf{v}(\mathbf{a}, t')}dt',$$

$$= \overline{\mathbf{v}}t, \tag{2.69}$$

where the averaging process is denoted by the over-bar. The dispersion of a fluid element about the mean streamline passing the locality **a** is found by subtracting (2.69) from (2.68)

$$\mathbf{x}(\mathbf{a}, t) - \overline{\mathbf{x}(\mathbf{a}, t)} = \int_0^t \{\mathbf{v}(\mathbf{a}, t') - \overline{\mathbf{v}}\}dt' = \int_0^t \mathbf{v}'(\mathbf{a}, t')dt',$$

where $\mathbf{v}'$ is the variation in velocity of the fluid element about its mean velocity as it moves along its trajectory. More concisely,

$$\mathbf{y}(\mathbf{a}, t) = \int_0^t \mathbf{v}'(\mathbf{a}, t')dt', \tag{2.70}$$

giving the displacement **y** of the fluid element relative to the mean position of the ensemble (the cloud) in terms of its velocity $\mathbf{v}'$ relative to the mean. The simplest measure of its dispersion is the second moment of the distribution, or the mean square value of **y**, and this can be expressed usefully in terms of the geometrical and flow characteristics.

In aquifer flows with random spatial variations in permeability, the direction of the mean streamline through a point is in the direction of the local pressure gradient, and the dispersal characteristics in the longitudinal, transverse and vertical direction may well be different (Hess *et al.*, 1992). In the longitudinal, 1-direction, say, the displacement of the fluid element relative to the centroid of the cloud is

$$y_1(\mathbf{a}, t) = \int_0^t v_1'(\mathbf{a}, t')dt',$$

or equivalently,

$$\frac{d}{dt}y_1(\mathbf{a}, t) = v_1'(\mathbf{a}, t), \qquad \text{with} \qquad y_1(\mathbf{a}, 0) = 0. \tag{2.71}$$

When the two left- and right-hand sides of these equations are multiplied together, there results

$$y_1\frac{dy_1}{dt} = \frac{1}{2}\frac{d}{dt}y_1^2 = \int_0^t v_1'(\mathbf{a}, t)v_1'(\mathbf{a}, t')dt'.$$

Upon averaging over many fluid elements from the same initial resolution volume, we have

$$\frac{d\sigma_1^2}{dt} = \frac{d}{dt}\left(\overline{y_1^2}\right) = 2\int_0^t \overline{v_1'(\mathbf{a}, t)v_1'(\mathbf{a}, t+\tau)}d\tau, \tag{2.72}$$

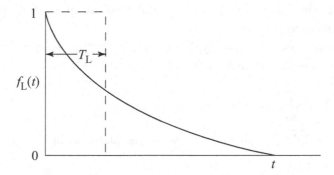

Figure 2.9. The Lagrangian autocorrelation function $f_L(\tau)$ decreases from unity when $\tau = 0$ and its integral is assumed to converge to the integral time scale T_L, which is defined as the width of the rectangle of equal area.

where τ is the time interval between the two velocity variations (relative to the mean) in the covariance of the integrand (Figure 2.9) and $\sigma_1^2 = \overline{y_1^2}$ is the variance or mean square displacement in the x-direction about the centroid of the marked fluid.

The velocity fluctuations of a marked fluid element are statistically steady in time, so that this covariance depends on the time interval τ, rather than on t and t' separately. When $\tau = 0$, the values of v_1' in the covariance are taken simultaneously, and it reduces simply to the mean square of the interstitial velocity component in the longitudinal direction, $\overline{v_1'^2}$. As the time interval increases, the interstitial velocity variation at the end of the interval begins to differ in a random way from what it was at the beginning as each element moves through the pores or along the interstices. The velocity variations relative to the mean become less correlated – they lose memory – and the covariance function decreases. Ultimately, when the interval $\tau - t$ is large, the velocity variation at the later time is assumed to have lost all dependence on what it happened to be initially. If this is so, the covariance function drops to zero as $\tau \to \infty$, in the manner shown qualitatively in Figure 2.9 and the integral is assumed to converge.

The covariance can be expressed equivalently as

$$\overline{v_1'(t)v_1'(t + \tau)} = \overline{v_1'^2} f_L(\tau) \tag{2.73}$$

where $f_L(\tau)$ is the dimensionless longitudinal Lagrangian correlation function (following the fluid elements) which has the value of unity when $\tau = 0$ (the fluctuation correlates perfectly with itself) and is assumed to decrease to zero (no correlation) as the time interval becomes very large. The integral time scale T_L, the area under the curve in Figure 2.9, is a useful measure of the time interval over which the correlation persists.

The rate of increase of the longitudinal variance of the cloud of fluid elements in the x-direction is therefore given from (2.71) and (2.72) as

$$\frac{d\overline{y_1^2}}{dt} = 2\overline{v'_1}^2 \int_0^t f_L(\tau)d\tau$$

$$\rightarrow 2\overline{v_1'^2} \int_0^\infty f_L(\tau)d\tau \qquad \text{when the elapsed time} \qquad \tau \gg T_L,$$

$$= 2\overline{v_1'^2}T_L, \tag{2.74}$$

where

$$T_L = \int_0^\infty f_L(\tau)d\tau$$

is the integral time scale of the longitudinal velocity fluctuations following the motion. Thus

$$\overline{y_1^2} \rightarrow \left(2\overline{v_1'^2}T_L\right)t = 2D_Xt, \quad \text{say,} \tag{2.75}$$

and $D_X = (2\overline{v_X'^2}\lambda_T)$ is the longitudinal macroscopic dispersion coefficient, with physical dimensions $[L^2T^{-1}]$. Asymptotically, the variance in length of the marked cloud increases linearly in time, and its average root mean square spread increases as $t^{1/2}$, factors common to most dispersive phenomena. The variances in lateral and vertical spread of marked fluid can be expressed in the same form, and involve the mean square lateral and vertical interstitial fluid velocities and the corresponding integral time scales.

Under natural field conditions the locally averaged permeability may vary continuously on the scale of the flow domain itself, producing corresponding velocity correlations that, albeit small, persist over considerable spatial intervals. Integrals based on measured data may not converge and the classical description above, though conceptually important, is less useful. An alternative approach using structure functions or "variograms" that are insensitive to large-scale variations in medium properties is described in Section 3.3.

2.10.2 Dispersion in a steady plume

In many applications the *spatial distribution* of contaminants from a discharge may be of greater interest than their precise time of arrival, and the results of the analysis above can be re-phrased to address this. Note that the Lagrangian analysis above is asymptotically exact within its defining parameters – a homogeneous, statistically steady, random velocity field, however it is generated. It contains no dynamics and makes no use of the defining property of interstitial fluid flow that the geometry of the interstices does not change in time. A parallel Eulerian version argument is only approximate, but does explicitly recognize this fact.

Consider a statistically steady plume of marked fluid arising from a local injection point in a statistically homogeneous aquifer. The total time derivative in (2.72) reduces to the advective rate of change $\overline{\mathbf{v}} \cdot \nabla$, and in the time interval $d\tau$, fluid elements have moved an *average* distance $dr_1 = \overline{v}_1 d\tau$, where $\overline{v}_1$ is the mean interstitial fluid velocity in the 1-direction or mean flow direction, so that in terms of the *spatial* covariance of the local interstitial velocity variations on scales of l_{AV} or greater,

$$R_{ij}(\mathbf{r}) = \overline{v'_i, (\mathbf{a})v'_j(\mathbf{a} + \mathbf{r})} \tag{2.76}$$

we have

$$\overline{v}_1 \frac{d}{dx_1}\left(\overline{y_1^2}\right) \approx 2 \int_0^{x_1} \overline{v'_1(a)v'_1(\mathbf{a} + r_1)}dr_1/\overline{v} = 2(\overline{v}_1)^{-1} \int_0^{x_1} R_{11}(r_1)dr_1,$$

Asymptotically,

$$\frac{d}{dx_1}\left(\overline{y_1^2}\right) \to 2\left(\overline{v_1'^2}/\overline{v}^2\right) \int_0^{\infty} f_{11}(r_1)dr_1 = 2\left(\overline{v_1'^2}/\overline{v}^2\right)\lambda_{11-1}, \tag{2.77}$$

where $f_{11}(r_1)$ is the dimensionless correlation function associated with the covariance (2.76) and λ_{11-1} is the integral length scale of the velocity variations in the 1-direction along the direction of flow as in Figure 3.3 in the next chapter. The mean square longitudinal dispersion of fluid elements therefore *asymptotically* increases linearly with travel distance

$$\left(\overline{y_1^2}\right) \to 2\left(\overline{v_1'^2}/\overline{v}^2\right)(\lambda_{11-1})x_1 = 2\left(\overline{u_1^2}/U^2\right)(\lambda_{11-1})x_1, \tag{2.78}$$

where u_1 represents the local variation in velocity about the mean U and x_1 is the distance from the source. Similarly, the lateral dispersion is given by

$$\left(\overline{y_2^2}\right) \to 2\left(\overline{u_2^2}/U^2\right)(\lambda_{22-1})x_1, \tag{2.79}$$

where λ_{22-1} is the integral length scale in the 1-direction of the velocity variations in the lateral 2-direction. The vertical dispersion about the mean streamline is expressed similarly, with 3 replacing 2 in (2.79). Since the variations in mean interstitial fluid velocity are produced by spatial variations in permeability, they are also proportional to the local *mean* velocity, and the first factor on the right of these expressions is purely numerical, generally smaller than unity, independent of the mean flow speed and dependent only on the local permeability structure.

The expressions

$$\alpha_{D-1} = \left(\overline{u_1^2}/U^2\right)\lambda_{11-1} \quad \alpha_{D-2} = \left(\overline{u_2^2}/U^2\right)\lambda_{22-1} \quad \alpha_{D-3} = \left(\overline{u_3^2}/U^2\right)\lambda_{33-1}$$

$$(2.80)$$

have the physical dimensions of [length], and are called the longitudinal, transverse and vertical *dispersivities*. They are the analogues of the macroscopic dispersion coefficients such as D in equation (2.75) that specify the spreading in time of a tagged fluid patch. Since the velocity ratios in (2.80) depend upon the permeability structure but are independent of the mean flow speed, the dispersivities are properties of the medium, not of the flow. Thus, from (2.78) and (2.79), the asymptotic root mean square spread of marked fluid in longitudinal and lateral directions increases with distance x_1 from the source as $(\alpha_{D-1}x_1)^{1/2}$ and $(\alpha_{D-2}x_1)^{1/2}$ etc.

3

Patterns of flow

3.1 Flow in uniform permeable media

Typical patterns of flow in a saturated aquifer are qualitatively different from those in a stream or lake. In the latter, flows are governed by a balance between gravity and fluid inertia, they may be vortical with closed circulation paths or eddies that range in scale from centimeters or less in a turbulent patch to the whole length of the lake. In an aquifer, the flow is governed by a balance between the gravitationally induced pressure gradients (or the internal buoyancy) on the one hand, and the flow resistance of the medium on the other. Fluid makes its way from recharge areas to discharge. We show below that circulating flows in a horizontal plane are impossible and that closed vertical circulation can be driven only by buoyancy variations, either positive or negative.

These flows are governed by the Darcy force balance (2.24):

$$\mathbf{u} = \frac{k}{\nu}(-\nabla(p/\rho_0) + b\mathbf{l}), \quad \text{where the buoyancy} \quad b = g\left(\frac{\rho_0 - \rho}{\rho_0}\right), \quad (3.1)$$

p is the reduced pressure, and $\mathbf{l}$ is a unit vector vertically upward. This is, in component form with axes (x, y, z) and velocity components (u, v, w) with z and w being in the vertical direction,

$$u = -\frac{k}{\nu}\frac{\partial(p/\rho_0)}{\partial x},$$

$$v = -\frac{k}{\nu}\frac{\partial(p/\rho_0)}{\partial y},$$

$$w = -\frac{k}{\nu}\left\{\frac{\partial(p/\rho_0)}{\partial z} + b\right\}, \quad (3.2)$$

where the permeability k and kinematic viscosity ν are here assumed to be constant. The Cartesian form of the incompressibility condition $\nabla \cdot \mathbf{u} = 0$ is

$$\frac{\partial u}{\partial x} + \frac{\partial v}{\partial y} + \frac{\partial w}{\partial z} = 0. \tag{3.3}$$

Note the structure of these equations. The buoyancy, associated with variations in temperature and/or salinity, is a field variable determined by the conservation equations (2.48) and (2.51), together with an appropriate equation of state. The buoyancy appears *only* in the vertical force balance. A local region of positive buoyancy drives the interstitial fluid vertically upward, yet the motion is generally fully three-dimensional. Coupling with the horizontal motion occurs through the incompressibility condition (3.3). As the buoyant region moves upward, adjacent fluid moves inward near the bottom to replace the fluid driven vertically and outward near the top.

3.1.1 Flow constraints

In the absence of buoyancy variations, the range of possible flow solutions in a uniform medium is very tightly constrained. The Darcy equation reduces to its classical form

$$\mathbf{u} = -\frac{k}{\mu}\nabla p. \tag{3.4}$$

Note that both the permeability k and viscosity $\mu = \rho_0 \nu$ are positive and the minus sign indicates that the transport velocity is always *down the gradient of reduced pressure*. An immediate consequence of this is that:

(i)   in constant density Darcy flow, no streamline can form a closed loop.

The proof of this statement by *reductio ad absurdum* is simple and quite general. Let us assume that it is not true, that there is indeed a closed streamline, and we find that this produces a contradiction. Start at some point on the streamline and move in the direction of flow; the reduced pressure would continually decrease, and as we complete the circuit and return to the same point, it would be less than it was at the beginning. But this is not possible since the pressure is a single-valued function, so that the assumption of a closed streamline must be false. Consequently, there cannot be any closed streamlines in a constant density Darcy flow, either in two or three dimensions, regardless of the non-uniformity or distribution of permeability.

A dynamical characteristic of permeable-medium flow pattern, second only in importance to the transport velocity is its curl, the *rotation vector*,

$$\Omega = (\Omega_x, \Omega_y, \Omega_z) = \nabla \times \mathbf{u} = \text{curl}(\mathbf{u}). \tag{3.5}$$

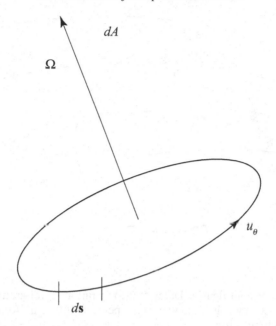

Figure 3.1. Illustrating Stokes' theorem, equation (3.6).

The rotation vector has the same appearance as the vorticity in ordinary fluid mechanics, but because it appears in the context of Darcy flow, it does not have the important dynamical properties possessed by the vorticity. It is, nevertheless, a useful property of flows driven by buoyancy and those in media of variable permeability, considered later in this book. If the permeability is uniform, the curl of the Darcy equation (3.1) is

$$\Omega = \nabla \times \mathbf{u} = \frac{k}{\nu} \{\nabla \times (b\mathbf{l})\}.$$

Since $\mathbf{l}$ is a unit vector vertically upward, the *vertical* component of Ω vanishes, so that

(ii)   for Darcy flow with variable buoyancy in a uniform medium, the *horizontal* flow field is irrotational.

A useful mathematical theorem due to Stokes provides an association between the rotation vector and transport flow circulation, as illustrated in Figure 3.1. This theorem (see Lighthill, 1966, pp. 55–57, or any good book on vector calculus) states that for *any* vector field $\mathbf{u}(\mathbf{x})$, the line integral around an arbitrary closed circuit (not a streamline circuit, because none exist) of the component of $\mathbf{u}(\mathbf{x})$ in the direction of the circumference, is equal to the surface integral of the normal

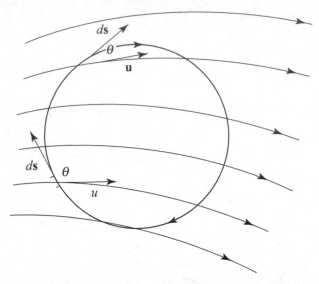

Figure 3.2. In constant density Darcy flow, the line integral around *any* closed loop of $\mathbf{u} \cdot d\mathbf{s}/k$ vanishes; in particular, if the permeability is uniform, the rotation vector is everywhere zero.

component of the rotation vector $\Omega = \nabla \times \mathbf{u} = \mathrm{curl}\,\mathbf{u}$ over any surface capping the circuit

$$\oint \mathbf{u} \cdot d\mathbf{s} = \int \nabla \times \mathbf{u} \cdot dA = \int \Omega \cdot dA, \qquad (3.6)$$

where $d\mathbf{s}$ is a differential element of the loop, a vector whose magnitude is the length of the element and whose direction is always that of the tangent to the loop in a consistent sense and dA is an element of area of the cap. With a small *circular* contour of radius r, the left-hand side of (3.6) is equal to the mean azimuthal speed u times the circumference $2\pi r$, and the right-hand side is equal to the mean rotation normal to the circuit times the area πr^2. Thus $\pi r^2 \Omega = 2\pi r u$ and so $\Omega = 2u/r$. The angular velocity associated with the distribution of transport velocity is u/r, or, from the last equation, $\frac{1}{2}\Omega$. The rotation can thus be visualized as twice the local angular velocity associated with the transport velocity field.

Further dynamical constraints can be established by considering the flow at points along a closed loop in a constant-density flow region, as illustrated in Figure 3.2. This loop cannot be entirely along a streamline in view of the proof above, but must cut across streamlines over at least some part of the loop. Consider (3.4) in the form

$$\frac{\mu}{k}\mathbf{u} = -\nabla p,$$

and integrate this around the closed loop. Notice from Figure 3.2 that $\mathbf{u} \cdot d\mathbf{s} = u(ds) \cos \theta$ is positive around the top of the loop since $\theta < \pi/2$, but negative around the bottom where the angle θ is obtuse, so that contributions to the integral around the whole loop include both positive and negative parts, which tend to cancel out. In fact, they cancel out *exactly*, even if the permeability varies along the path. The line integral around the complete loop is

$$\mu \oint \frac{\mathbf{u} \cdot d\mathbf{s}}{k} = - \oint \nabla p \cdot d\mathbf{s} = p_1 - p_2,$$

which vanishes because the circuit is closed, the finishing point 2 being the same as the starting point 1, and the pressure is single-valued. Thus,

(iii)   even though the medium permeability k varies spatially (with the texture or composition of the matrix), in the absence of buoyancy variations the integral around *any* closed loop of the tangential velocity divided by the permeability vanishes:

$$\oint \frac{\mathbf{u} \cdot d\mathbf{s}}{k(\mathbf{x})} = 0. \tag{3.7}$$

This is already a useful expression. Note that it is consistent with the flow boundary condition at an internal interface described in Section 2.4, i.e. that the ratio of the tangential velocity components on each side is equal to the ratio of the permeabilities, as the reader may verify by taking a thin loop that runs along one side of the boundary a distance $d\mathbf{s}$, then across and back on the other side and across to the starting point.

From Stokes' theorem, equation (3.7) is equivalent to

$$\nabla \times \left\{ \frac{\mathbf{u}(\mathbf{x})}{k(\mathbf{x})} \right\} = 0, \tag{3.8}$$

whence, by writing out the components of the curl,

$$\Omega = \nabla \times \mathbf{u} = \nabla \left(\ln \frac{k(\mathbf{x})}{k_0} \right) \times \mathbf{u}, \tag{3.9}$$

where Ω is the rotation vector of the transport velocity field and k_0 is any convenient reference value of the permeability field. Note that if a different reference value is chosen, the logarithm is different by an additive constant, which is immaterial since the permeability appears only in the gradient of the logarithmic term. The rotation vector theorem (3.9) demonstrates that *variations in log-permeability generate rotation in an otherwise uniform stream* and provides a general but direct connection between the velocity and permeability distributions without specific reference to the pressure distribution. It asserts, for example, that in a largely horizontal flow, streamlines cutting *across* a log-permeability gradient induce rotation about the vertical axis, attracting streamlines to regions of high permeability and repelling them from regions of low permeability. It is used in Section 3.3 concerning fluid flow dispersion in an extensive aquifer with spatially random permeability.

In a homogeneous region of constant density and permeability, (3.7) becomes

$$\oint \mathbf{u} \cdot d\mathbf{s} = 0, \tag{3.10}$$

so that, from (3.6) or (3.9), all components of the rotation vector vanish, $\Omega = 0$. Thus,

(iv)   constant-density Darcy flow in a homogeneous, uniform medium is everywhere irrotational, with the circulation around *any* closed loop vanishing, and the rotation vector everywhere zero.

This is a stronger constraint than (ii) above. It can also be seen directly from (3.5), with $b = 0$. These statements (i)–(iv) limit substantially the range of dynamically possible flow patterns in uniform permeable media.

3.1.2 Laplace's equation

Let us return to the incompressibility and Darcy equations to show that a similar direct connection can be found between the permeability and pressure fields. The divergence of the Darcy equation (3.4) gives

$$\nabla \cdot \mathbf{u} = -\mu^{-1} \nabla \cdot (k\nabla p) = -\mu^{-1}(\nabla k \cdot \nabla p + k\nabla^2 p) = 0,$$

because the fluid is regarded as incompressible and divergence-free. On rearrangement, this becomes

$$\nabla^2 p + k^{-1}\nabla k \cdot \nabla p = 0$$

or

$$\nabla^2 p + \nabla \ln(k/k_0) \cdot \nabla p = 0, \tag{3.11}$$

where k_0 is a reference value for the permeability distribution, frequently taken as its geometric mean. Note that here, as well as in equation (3.9), the lack of homogeneity in the medium is expressed by the log-permeability, rather than the permeability itself. This equation specifies the distribution of pressure (and thence, transport velocity) in flow through an aquifer with a prescribed internal log-permeability distribution and appropriate pressure boundary conditions.

If the medium is *homogeneous*, the second term in (3.11) vanishes and it reduces simply to

$$\nabla^2 p = 0, \tag{3.12}$$

which is known as Laplace's equation. In Cartesian coordinates, it has the form

$$\left(\frac{\partial^2}{\partial x^2} + \frac{\partial^2}{\partial y^2} + \frac{\partial^2}{\partial z^2} \right) p = 0. \tag{3.13}$$

It is one of the most basic and best-studied differential equations in all of physics, with applications to field theories in electricity, magnetism, hydrodynamics, as well as here, in flow in permeable media. Conventionally, in fluid flow situations, the Cartesian x and y coordinates are taken in the horizontal plane with the z coordinate vertically upward. In two-dimensional flow, the equation above reduces to

$$\left(\frac{\partial^2}{\partial x^2} + \frac{\partial^2}{\partial y^2} \right) p = 0. \tag{3.14}$$

In a uniform medium, the stream function (2.13), which automatically satisfies the incompressibility condition, also satisfies Laplace's equation in two dimensions. For if $\mathbf{u} = (u, v)$ in the (x, y) directions, then since $u = \partial\Psi/\partial y$, $v = -\partial\Psi/\partial x$, and with use of Darcy's equation, we have

$$\frac{\partial^2\Psi}{\partial y^2} = \frac{\partial u}{\partial y} = -\frac{\partial}{\partial y}\left(\frac{k\partial p}{\mu\partial x} \right) = -\frac{\partial}{\partial x}\left(\frac{k\partial p}{\mu\partial y} \right) = \frac{\partial v}{\partial x} = -\frac{\partial^2\Psi}{\partial x^2},$$

so that

$$\frac{\partial^2\Psi}{\partial x^2} + \frac{\partial^2\Psi}{\partial y^2} = 0. \tag{3.15}$$

This is often a more convenient alternative to (3.14).

Many analytical and numerical methods are available for solving Laplace's equation with a variety of boundary conditions reflecting the variety of possible flow geometries. Once a solution for the pressure p or the stream function Ψ has been found, the transport velocity field (in our applications) can be found from (3.4) or the definitions (2.13), but a few general properties of all such solutions are worth keeping in mind. Among them are the following.

(i) The total fluid pressure $p_T = p + p_h = p - \rho g z$ (see Section 2.4) also satisfies Laplace's equation, since the hydrostatic pressure is linear in z and its second derivative vanishes.

(ii) If the solution p for the reduced pressure has no singularities, it cannot have a maximum in the interior of the flow domain. This can be seen most simply in two dimensions. For a function p in two variables, x and y, in geometrical terms, the second derivatives in the x and y directions are measures of its curvature in these directions. Where the curvature is positive, the slope is increasing, and where negative, the slope is decreasing. At an interior maximum, the slope in both directions at that point is zero and the curvatures in both directions are negative. This is inconsistent with Laplace's equation, (3.14), which asserts that the total curvature, the sum of the two principal curvatures, is zero. Locally, the two-dimensional solutions must be saddle-shaped. The solution cannot have an interior minimum either, for the same kind of reason. The pressure maxima and minima must lie at points on the *boundary* of the domain.

This conclusion can be seen in physical terms as follows, which applies in three dimensions as well. If there were a pressure maximum in the interior, the Darcy equation (3.1) implies that fluid is flowing away from the maximum in all directions. This violates the incompressibility condition, so that the assumption of an interior pressure maximum must be false. The same argument can be used for a minimum.

(iii) Singularities are of more than mathematical interest in aquifer hydrology, however. A pumped well in an aquifer constitutes a local singularity, a sink in the groundwater flow pattern where the water table does have an interior minimum, toward which the water flows. An injection well is a local source singularity where the water table is locally highest with flow diverging outward. As described above, two-dimensional solutions $\Psi(x, y)$ of Laplace's equation have curvatures that are either both zero or are equal and opposite in the two principal directions. In the stream function solution for flow surrounding a source singularity, at every point the curvature is negative in a radial plane (slope decreasing outwards), and positive in the vertical tangential plane (zero in the direction perpendicular to the radius, curving upward on both sides).

(iv) In two-dimensional solutions to Laplace's equation, the length scales over which the derivatives of p and Ψ (the velocities) vary in the two orthogonal directions, are the same. In three-dimensional solutions, the length scales in all three orthogonal directions are generally comparable; no one length scale can be much less than the other two. One may, however, be much greater than the other two, which are then comparable; this is the case in "two-dimensional" flow.

(v) As discussed in Section 2.4, solutions to Laplace's equation (3.12)–(3.14) are subject to conditions that define the nature of the flow boundaries, and a little care is sometimes required to ensure that these boundary conditions are complete and mutually consistent. Classical texts on potential theory and Laplace's equation show that, in general, either the solution variable (i.e. p or Ψ) or its normal derivative, but not both, must be prescribed at all points along the boundaries of the domain. Situations involving the water table have occasionally caused trouble in this field, since the water table configuration and the infiltration rate cannot both be prescribed. When doubt exists, it is probably best to use common sense, to imagine that a flow experiment is being done and decide which variables one can control, and which will be a consequence of the experiment. In experimenting on flow through a sand bed, one can control the rate and distribution of water sprinkled on the top of the sand, while the position and configuration of the water table inside the sand adjusts itself in the experiment. One can prescribe the pressure at the discharge (usually atmospheric) but not the distribution of flow. The mathematics should conform to these causalities.

3.1.3 Some local flow patterns

A number of simple solutions for local perturbations to a uniform flow illustrate the characteristics described above while also having geological interest. A fluid inclusion in an almost impermeable rock matrix and the flow around an abraded permeable rock in a sandy aquifer may share a common spherical geometry, with radius a, say, and permeability k_1 inside and k_0 outside. Let the undisturbed ambient pressure gradient be ϖ. Laplace's equation specifying the reduced pressure distribution in each region, has separate spherical harmonic solutions p_1 and p_0, as described in books on potential theory, which are connected through the following boundary conditions: (i) the internal and outside fluid pressures, p_1 and p_0, respectively, must be the same at all points on the surface $r = a$; (ii) the normal component of the transport velocity across the boundary, $u_r = -(k/\mu)\partial p/\partial r$ must also be the same on both sides; and (iii) far from the sphere the pressure distribution must approach the undisturbed uniform gradient, $-\varpi$, say. Thus,

$$p_1 = p_0, \qquad k_1 \frac{\partial p_1}{\partial r} = k_0 \frac{\partial p_0}{\partial r} \qquad \text{at} \quad r = a, \tag{3.16a}$$

and

$$p_0 \to -\varpi x = -\varpi r \cos \theta \qquad \text{as } r \to \infty, \tag{3.16b}$$

where r is the radial coordinate and θ is the angle between this and the direction of the external flow. The spherical harmonic solutions for the internal and external pressure distributions are

$$p_1(r, \theta) = -\varpi \frac{3k_0}{k_1 + 2k_0} r \cos \theta \qquad \text{inside the sphere} \tag{3.17a}$$

and

$$p_0(r, \theta) = -\varpi \left\{ 1 - \left(\frac{k_1 - k_0}{k_1 + 2k_0} \right) \frac{a^3}{r^3} \right\} r \cos \theta \qquad \text{outside.} \tag{3.17b}$$

Inside the sphere, the pressure (relative to that at the center) is proportional to $r \cos \theta = x$ so that for any permeability ratio, the interior flow u_1 is *uniform* and entirely in the x-direction. From the solution above, its magnitude is

$$u_1 = -\frac{k_1}{\mu} \frac{\partial p_1}{\partial x} = \varpi \frac{3k_1 k_0}{k_1 + 2k_0} = \varpi \frac{3k_0}{1 + 2(k_0/k_1)},$$

which differs from that in the ambient fluid by the factor $G = 3k_1/(k_1 + 3k_0)$. When the spherical region is less permeable than that outside, $k_1 < k_0$, this factor is smaller than 1 and the flow is partially deflected around it, in accordance with intuition. If it is *much* less permeable than the ambient as it is for a rock in a sandy aquifer, the factor is near zero and almost no fluid passes through, as

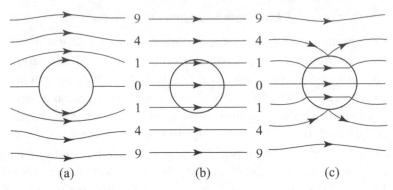

Figure 3.3. Stokes stream function solutions for the flow around and through spherical inclusions. In the first figure (a), the permeability is much less than in the surrounding matrix and the fluid flows around the inclusion; in the second (b), the permeability inside and outside are the same; and in the third (c), the inclusion is filled with fluid, which moves through the inclusion three times faster than in the undisturbed state.

shown in Figure 3.3. When $k_1 > k_0$, the inside is now *more* permeable and fluid is attracted to pass through the sphere. In the limit, when the interior is simply fluid filled, it can be considered infinitely permeable, so that (mathematically) $k_0/k_1 \rightarrow 0$ and the focusing factor $G = 3$ precisely; the fluid passing through the sphere moves just three times faster than the transport velocity in the undisturbed surroundings.

Exact analytical solutions can also be found for the two-dimensional flows attracted to thin isolated cracks of finite length or around similar cracks that have been filled with impermeable precipitate. They are simple modifications of solutions for potential flow in classical hydrodynamics given, for example, in Lamb's *Hydrodynamics* (1932, p. 86) and the salient flow properties are illustrated in Figure 3.4. If the ambient flow velocity U is parallel to the plane of a highly permeable crack of length l, say, and perpendicular to its leading edge, fluid is attracted to the crack from a depth range precisely equal to its length, passes through it and is re-injected into the ambient. The volume flux inside the crack is zero at the leading edge, increases to a maximum of Ul at the center, and returns to zero at the end. If the stream is oblique to the flow direction as in Figure 3.4b, the oncoming fluid turns towards the crack and moves some distance along it before re-entering the medium and turning back to its original flow direction. A relatively small amount of fluid passes beyond the leading edge before entering on the rear. If the oncoming stream is perpendicular to the plane of the crack, it simply passes through undeflected.

When the crack is sealed by deposition of minerals from solution (in Figure 2.4 for example), the flow patterns can contain stagnation points, where the streamlines

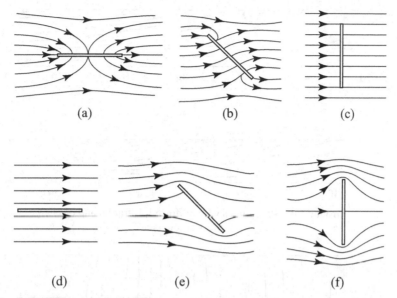

Figure 3.4. Exact two-dimensional solutions of Laplace's equations representing flow streamlines near thin lenses. In the upper three panels, the lens is much more permeable than the matrix or is a liquid-filled crack, and in the lower three, it is impermeable.

divide and pass around the obstacle in opposite directions, as in Figures 3.4e and f.

3.1.4 Two-dimensional surface aquifers

The exact solution to Laplace's equation in a layer of thickness d above an impermeable basement with a uniform infiltration rate W across the upper surface is

$$\Psi = (W/d)xz, \qquad x > 0, \tag{3.18}$$

as illustrated in the top panel of Figure 3.5. This provides a reasonable first order representation of the groundwater flow to the right of the groundwater divide in a uniform aquifer, but it has some shortcomings. In nature, the pressure gradient required to drive the interstitial fluid to the right is supplied by the slope of the groundwater table from its maximum elevation at the point of divide, so that the upper boundary of the flow is not, in fact, flat. Nevertheless, it represents accurately the flow pattern through the main body of a uniform two-dimensional aquifer, with streamlines that are equally spaced at each vertical section, indicating the uniformity of the transport velocity with depth, and their reduced spacing with distance from the groundwater divide indicating the increase of the horizontal flow with distance from that divide. However, the flow in a natural aquifer fluid ultimately discharges

Patterns of flow

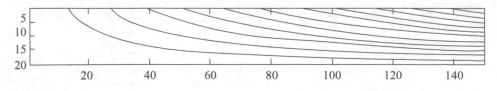

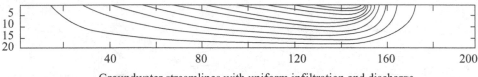

Groundwater streamlines with uniform infiltration and discharge
into a shallow lake

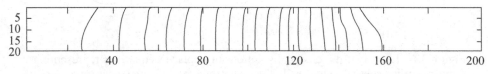

Reduced pressure contours

Figure 3.5. The upper panel represents the exact solution (3.18); the center panel shows streamlines in a two-dimensional aquifer with uniform infiltration from above in the range $0 < x < 150$ and constant-pressure discharge beyond (into a lake bed, for example). The lowest panel shows the isobars, the lines of constant reduced pressure, which slope forward near the groundwater divide where the streamlines trend downward, and backward near the discharge to drive the fluid upward.

by seepage upwards through a stream-bed and laterally through its banks, or through the near-shore bed of a lake or estuary, and the exact solution above does not reflect this.

The center panel shows streamlines from a numerical calculation in which the upper boundary representing the water table terminates in a swamp or shallow lake whose free surface to the atmosphere is horizontal. In this region, the upper boundary condition with a prescribed infiltration rate W is replaced by a condition of constant pressure, with fluid being free to discharge across the boundary into the body of water above. Notice that in this discharge region, the flow spreads horizontally beyond the shore line to a distance of the order of the aquifer depth, as is characteristic of two-dimensional solutions of Laplace's equation discussed earlier. The lower panel shows isobars, or lines of constant reduced pressure, with higher values to the left, beneath the water table maximum, and lower values in the discharge region. Beyond the immediate vicinity of the groundwater divide, the isobars are very nearly vertical but leaning slightly forward, consistent with the gradually deepening streamlines. In the central part of the aquifer, the flow is very nearly horizontal and the isobars are nearly vertical. As the discharge region is approached, the isobars begin to slope backward, the distribution of total pressure is no longer hydrostatic, the reduced pressure at depth begins to exceed that above, and the streamlines begin to rise towards the surface discharge, ultimately becoming vertical at the interface.

Natural surface aquifers are usually much more complex than this schematic model. Sedimentary deposits can be laid down as vertical sequences of approximately horizontal layers having sometimes highly contrasting permeability. For example, the entire Delmarva Peninsula between the Chesapeake Bay and the Western North Atlantic Ocean, sectioned schematically in Figure 3.6, consists of a series of permeable, sandy aquifers whose cementation increases with depth, interspersed by much less permeable "retarding layers" of finer clay or silt. The permeability of these is smaller than in the aquifers by factors of 10^{-3} to 10^{-4} (Shedlock *et al.*, 1999). Although the material in the individual layers may be locally isotropic in its flow properties, the sequence of layers clearly influences the larger-scale flow in a highly anisotropic way. Specifically, a structure of multiple layers with significantly different permeability channels the flow into the layers of high permeability and low flow resistance, thereby avoiding the low-permeability, high-flow-resistance regions. For a sufficient permeability contrast, the actual value of the lowest permeability may become almost irrelevant since there is very little flow there. In contrast, the topology or structural geometry of the region, specifically the connectivity among the individual aquifer layers, is of considerable importance, but is often not known well.

3.2 Three-dimensional surface aquifer flow

This section is concerned with the overall fluid dynamics of aquifer flows. The next few pages are mostly qualitative, with simple physical descriptions and explanations to introduce the phenomena and to prepare for the more quantitative discussion that follows.

3.2.1 How do surface aquifers work?

Following a wet spell, water infiltrates through the unsaturated vadose zone and into the water-saturated region below. Near the groundwater divide, the water table elevation increases, but subsequently the horizontal pressure gradients relax with spreading flow in continuous, near-hydrostatic adjustment. In spite of the smaller permeability of any near-surface "confining layers," the very slow downward vertical flow requires a vertical pressure gradient only slightly less than hydrostatic.

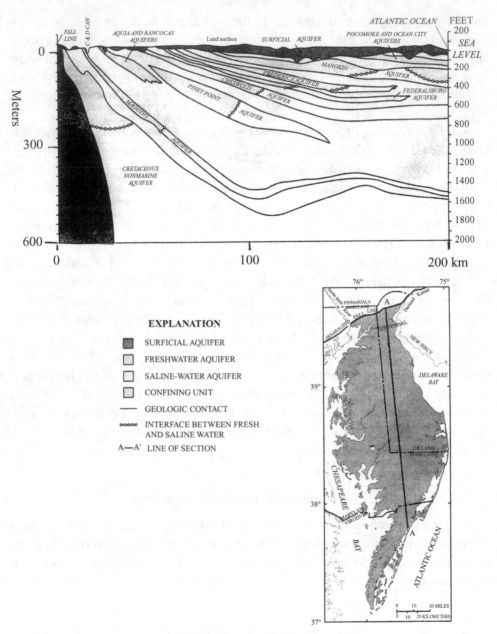

Figure 3.6. A schematic north/south section of the aquifer structure of the Delmarva Peninsula, from Shedlock *et al.* (1999).

In those flow tubes reaching to increasingly greater depths, the accumulating pressure deficit (below hydrostatic) required for downward flow increasingly subtracts from the pressure needed to move the groundwater along the aquifer toward discharge, and the flow in these tubes begins to stagnate. In spite of the expected complexity of realistic structural geometry, then, there usually does exist an "effective hydraulic depth," above which the horizontal pressure gradient is still close to that produced by the hydrostatic head of the water table vertically above, and below which it becomes increasingly smaller.

As the fluid moves downward from the water table, individual fluid elements are displaced by more recent infiltration from above, so that the time interval since they crossed the water table and lost contact with the atmosphere (the "groundwater age" of the fluid elements) continually increases with depth. The oldest groundwater is found throughout an aquifer at points close to the effective hydraulic depth, as indicated by the analysis and measurements to be described in more detail below. This remarkable property continues to be generally true no matter where the water column migrates; if it moves to a region where the aquifer is thinner or thicker, the pattern is compressed or stretched like a concertina but the general ordering of the groundwater ages remains the same. It does, however, depend to some extent on the uniformity in aquifer permeability. The flow is predominantly in the more permeable aquifer components but the few fluid elements that do drift into a less permeable, retarding layer tend to stay there for a relatively long time, so that some fluid elements from relatively shallow depth may be anomalously old. There is some evidence for this in the field measurements shown in Figure 3.10, below.

The progression of fluid elements through the aquifer is clearly of interest in questions of contaminant dispersal. A fluid element, crossing the water table, moves downward and is advected (transported largely horizontally) in the general direction of discharge by the mean groundwater stream that has been supplied by prior infiltration upstream. It may also be dispersed by the small-scale inhomogeneity of the medium. As the element moves downstream, its mean advection velocity increases over the length of the aquifer because of the increasing total infiltration along the lengthening flow path behind it – it accelerates constantly and drifts deeper as it moves towards discharge. This effect can be seen in the schematic streamline pattern in the central region of Figure 3.5. The uniformity of the streamline spacing at any section reflects the uniform vertical profile of the downstream velocity, while the convergence of streamlines reflects the longitudinal acceleration of the fluid particles.

Packets of fluid elements that may have been marked by a dynamically passive contaminant (one that does not affect the flow) are advected by the mean streaming, distorted by the internal strain field and dispersed by the random geometry of the internal flow paths and small-scale structural inhomogeneity. In a horizontally

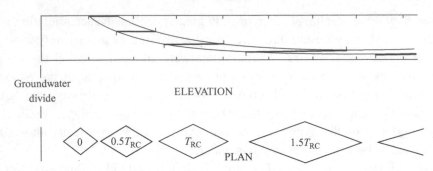

Figure 3.7. A diamond-shaped patch of marked fluid enters a two-dimensional aquifer across the water table near the groundwater divide and its subsequent positions and shapes are shown after elapsed multiples of 0.5 of the recharge time: $T_{RC} = \phi d / \overline{W}$. Note the acceleration and the longitudinal stretching in this region as the patch moves toward the basement.

uniform aquifer, the vertical component of the flow originating from infiltration decreases monotonically with depth from its climatic value in the flow across the water table, towards zero at the effective hydraulic depth. Shallower fluid elements sink more rapidly than deeper ones. Vertical pairs of marked fluid elements *converge* as they move downward at a rate $\partial w / \partial z$ determined by the local infiltration rate and the local aquifer thickness. Because of incompressibility, this vertical convergence must be accompanied by a *net horizontal divergence*, $\partial u / \partial x + \partial v / \partial y$, in the orthogonal directions. As patches of contamination move through the aquifer over time scales of possibly decades, they become flattened, thinner, more extended longitudinally and sometimes laterally also, as they drift downward and toward the discharge. This is illustrated in Figure 3.7. The total horizontal divergence is determined by the local infiltration, but the directional partition of it, along the lines of flow and transverse to them, is *not* determined locally, but by the shape of the possibly distant aquifer boundaries from groundwater divide to lines of discharge. Where flow lines are converging towards discharge, as can be seen in some areas in the upper part of Figure 3.9 below, the longitudinal divergence is large enough to overcome the lateral convergence, so that the net horizontal divergence remains positive.

Some natural geometries are fairly simple. For example, Long Island, off the coast of Connecticut, is a long, sandy aquifer in which the predominant groundwater flow is lateral, towards Long Island Sound on the north-west side and towards the Atlantic ocean on the south-east. As a first approximation, the flow field might be considered two-dimensional in transverse north-west to south-east sections. With (x, y) coordinates, the groundwater transport velocity v in the longitudinal direction is very small, and the motion is essentially two-dimensional. The horizontal flow divergence $\partial u / \partial x$ alone balances the infiltration rate. On the other hand, in a circular island aquifer with uniform infiltration and discharge around the circumference, the water table contours are circular. A circular patch of marked fluid remains circular, but expands as it moves toward the perimeter, so that the horizontal divergence is isotropic with equal longitudinal and transverse components. Most aquifer geometries are of course more complex than these.

3.2.2 Regional scale aquifer flow

In this section, we begin to take a more practical and quantitative approach. Consider a surface aquifer or a system of aquifers in which the overall aspect ratio is large, i.e. where is the typical length l of a flow path from groundwater divide to discharge is very much larger than the effective hydraulic depth d, and where the surface relief and the dip of embedded layers are both relatively gentle. A fully three-dimensional calculation of the groundwater flow in a particular aquifer system would require a detailed knowledge and specification of the structure of internal conduits, lenses and layers that is not usually available, but a more modest alternative *is* attainable that allows significant comparison with various kinds of field measurements. The gentle relief, thin layer approximation is a powerful and accurate simplification of the flow equations applicable to many aquifer situations. It concentrates our attention on *vertically integrated* properties of the flow such as the total flux distribution and distribution of groundwater levels, which do not require detailed information on the internal structural geology. A basic parameter in this approach is then the transmissivity, or vertically integrated hydraulic conductivity, which in a given aquifer can be estimated with adequate accuracy from a little theory and simple historical observations.

First, the incompressibility condition for the interstitial water (in effect, water volume conservation) can be applied to a vertical column of the medium of unit cross-section. It expresses the balance among surface infiltration from rainfall (W), the rate of increase in fluid volume occasioned by the vertical rise of the water table at the rate $\partial \zeta / \partial t = \dot{\zeta}$ and the lateral volume flux divergence. The total volume flux is the vertical integral of the horizontal transport velocity from the effective hydraulic basement at depth d upward to the water table:

$$\mathbf{q}(x, y) = \int_{-d}^{\zeta} \mathbf{u}(x, y, z, t)dz. \qquad (3.19)$$

The divergence of this flux $\nabla \cdot \mathbf{q}$ represents the net outflow from the sides of the column, so that the balance described above is represented by

$$W(t) = \phi \frac{\partial \zeta}{\partial t} + \nabla \cdot \mathbf{q}, \qquad (3.20)$$

where W is the rate of infiltration across the water table, taken positive, and ϕ is the porosity. This can be derived more formally by a vertical integration of the incompressibility condition $\nabla \cdot \mathbf{u} = 0$ with the boundary condition (2.28) and (2.30) at the water table and the condition of no flux across the basement, $w(-d) = u_h(-d) \cdot \nabla_h(-d)$.

To the thin layer approximation for the aquifer recharge regions, the total fluid pressure is very nearly locally hydrostatic:

$$p = \rho g(\zeta(x, y, t)). \tag{3.21}$$

The *horizontal* component of flow is driven by the horizontal gradient of the pressure:

$$\mathbf{u}(\mathbf{x}, z, t) = -(k/\mu)\nabla p = -K(z)\nabla_h\zeta(\mathbf{x}, t), \tag{3.22}$$

where $\mathbf{x} = (x, y)$ is horizontal position and in a horizontally layered aquifer, the hydraulic conductivity $K = gk/\nu$ may be a highly variable function of depth but a much more gradually varying function of horizontal position. Note the very important fact that the horizontal and vertical space variables $\mathbf{x}$ and $\mathbf{z}$ have separated in (3.22), A vertical integration of this equation and use of (3.19) gives

$$\mathbf{q} = -\nabla_h\zeta \int_{-d}^{\zeta} K(z)dz,$$

$$= -C\nabla_h\zeta, \qquad \text{say,} \tag{3.23}$$

where

$$C = \int_{-d}^{\zeta} K(z)dz \tag{3.24}$$

is the vertically integrated hydraulic conductivity of the aquifer, called the *transmissivity*. The transmissivity C may vary on a basin-wide scale, though in the applications considered here, it is generally considered constant, largely because the simple observations used to estimate it provide not point values, but a local average. A most important observation is that the dependence on depth has disappeared from the governing equations (3.20) and (3.23).

These two can be combined into a single equation by substitution of (3.23) into (3.20):

$$\phi\frac{\partial\zeta}{\partial t} = \nabla \cdot \{C\nabla\zeta\} + W(t), \tag{3.25}$$

which is a slight generalization of the classical two-dimensional diffusion or heat conduction equation with a possibly variable "diffusivity" C which here is the transmissivity, and a distributed infiltration source $W(t)$ that must be provided as input data. For a given infiltration history, $W(t)$, this is to be solved for $\zeta(\mathbf{x}, t)$, the water table elevation in horizontal space x, y and time t. The spatial distribution of total flux in the aquifer $\mathbf{q}(\mathbf{x}, t)$ is then found by substituting this solution into (3.23).

If the detailed vertical distribution of hydraulic conductivity $K(z)$ is known at one or more locations as a result of core measurements (or is conjectured) the right-hand side of (3.22) is determined completely, and the *vertical distribution* of flow velocity at that location can be recovered. This procedure reduces the three-dimensional, time dependent calculation to a *two*-dimensional, time-dependent calculation, with a great increase in computational efficiency.

When the geometry is particularly simple, analytical solutions for the mean water table elevation ζ can be obtained from (3.25). For example, in an idealized peninsula or isthmus with uniform C and constant width $2l_A$ the aquifer flow is unidirectional and in the x-direction. In the mean, or in a steady state, equation (3.25) reduces to

$$0 = C\frac{d^2\zeta}{dx^2} + W,$$

with $\zeta = 0$ at $x = \pm l_A$. The solution, specifying the distribution of water table height ζ above the surrounding mean water level is given by

$$\zeta = \frac{l_A^2 \overline{W}}{2C}\left\{1 - \left(\frac{x}{l_A}\right)^2\right\} = \tfrac{1}{2}h_W(1 - (x/l_A)^2), \qquad -l_A < x < l_A. \quad (3.26)$$

This is sometimes called the Dupuit–Forchheimer solution. The maximum elevation at the centerline above datum (i.e. the discharge level) is proportional to $h_W = l_A^2 \overline{W}/C$, where l_A is the path length of the aquifer flow. The height h_W is a natural scale for differences in aquifer water table level, while the factor $1/2$ and the precise parabolic profile in x are properties of this particular geometry. The groundwater transport velocity u increases linearly with distance from the groundwater divide, since

$$u = -K\frac{\partial \zeta}{\partial x} = \frac{K\overline{W}}{C}x, \qquad \text{from the above,}$$

$$= \frac{\overline{W}x}{d}, \qquad\qquad\qquad\qquad (3.27)$$

where d is the effective hydraulic depth. In this example with unidirectional flow, at every section x, the prior upstream infiltration $\overline{W}x$ is balanced by the outflow ud at the section.

A similar expression can be found for a circular island in which x now represents radial position. The radial transport velocity also increases linearly from the center to the circumference. The total infiltration over the catchment area increases quadratically with radius, and to balance this, the circumference and the efflux per unit length of circumference both increase linearly with radius. The numerical factor involved in h_W now turns out to be (1/4) rather than (1/2). It is again of order unity, but is smaller in magnitude than in the previous example, since radial advection in all directions spreads more rapidly than purely lateral advection in one.

3.2.3   An example: the aquifer in Kent County, Maryland

The simplicity of the vertically integrated aquifer flow equations (3.25) and (3.23) allows for the development of efficient numerical solutions for regional flows with realistic aquifer configurations. They are, of course, subject to appropriate initial and boundary conditions that characterize the domain. Along a running stream where groundwater discharges upward from the river bed and from the banks just above the water surface, the water table is close to the mean water level of the stream (or a meter or so above). As seen earlier, discharge regions are usually very local, with length scales of the order of the aquifer thickness. When the groundwater discharges to a marsh or coastline or shallow lake, the deep streamlines in the aquifer begin to turn upward as the horizontal pressure gradient provided by the water table slope begins to weaken over a distance proportional to the aquifer thickness. This is evident in the lower panel of Figure 3.5. The water level in the lake is of course horizontal, but the increased pressure (above hydrostatic) needed to force the water upward comes from the higher water table level just inland. The discharge from the entire depth of the aquifer emerges over a distance offshore that is again of the order of the aquifer thickness. This kind of near-shore, bottom discharge of groundwater has been observed, notably by Tokunaga *et al.* (2002) near the mouth of the Kurobe River, Japan, and has been simulated numerically by Wilson (2005) and by Prieto and Destouni (2005). In the case of discharge at a stream-bed draining the aquifer, the groundwater streamlines converge from both sides, throughout the whole depth of the aquifer. The strong upward velocity at the discharge is associated with relatively large near-surface hydraulic gradients that have been measured in the Delmarva Peninsula by Böhlke and Denver (1995, p. 2327). Apparently, these general flow patterns are fairly characteristic in natural aquifer shoreline situations, though any strong internal layering can be expected to stretch the horizontal scales somewhat.

In a regional flow calculation, it is assumed that the water table elevation ζ at any point cannot exceed the height ζ_S of the surface topography, which, like $W(t)$ must be provided as input data. This constraint anchors the water table level to coincide with, or be a little above, the surface of an accumulating, running stream along which discharge can occur. On the scale of the aquifer, the solution will have discontinuities in slope along these discharge lines, with the water table sloping upward from the stream as groundwater flux enters from one or both sides. The discharge flow is significantly non-hydrostatic, but is restricted to boundary regions whose width is again of the order of the aquifer thickness, at most. In a typical numerical calculation of regional hydrological flow, this is usually less than the horizontal grid spacing. The non-hydrostatic discharge region is therefore a sub-grid scale phenomenon, parametrized by this constraint or inner discharge boundary condition, which is, in fact, a principal determinant of the solution. In an unsteady time-stepping calculation, the anchoring can be implemented simply by an instruction at each step equivalent to:

$$\text{"if at any point, } \zeta(x, y, t) > \zeta_S(x, y) \text{ then put } \zeta(x, y, t) = \zeta_S(x, y)\text{"}. \quad (3.28)$$

Equation (3.25), with this constraint, allows calculation of the space-time evolution of the water table configuration, given the infiltration history and the transmissivity of the aquifer. In the mean, or in a steady state, equation (3.25) reduces to a balance between infiltration and lateral diffusion that allows a simple geometrical interpretation. When the regional variations in transmissivity C are small, $\nabla^2 \zeta = -\overline{W}/C$, where $\overline{W}$ is the mean infiltration rate, so that the water table curvature in an infiltration region is uniform and negative, i.e. convex upward. The water table therefore has the general form of flattened domes or barrel vaulting between downwardly pointing cusps along the lines of discharge. Once the distribution of water table elevation has been found, the pattern and magnitudes of the total fluxes of groundwater are specified from (3.23) above.

This kind of analysis has been applied by the present author (Phillips, 2003) to a calculation of the water table configurations and groundwater flow patterns in a 13.5-km square region of the surface aquifer between the tidal Chester and Sassafras Rivers in the Delmarva Peninsula, Maryland. The aquifer in this region is fairly uniform with porosity $\phi \sim 0.3$, about 20 m thick and underlain by a thick confining layer. The surface topography in the center of the region is essentially flat but is intersected by natural creek beds a few meters below the general land level. These flatten out towards the Chester River to the south. The northern part of the region along the Sassafras River is generally higher, indented by small bays and with shorter drainage streams. The infiltration history $W(t)$ was inferred by Reilly *et al.* (1994) from estimates of the average infiltration fraction and historical rainfall records from Chestertown, near the south-west corner of the

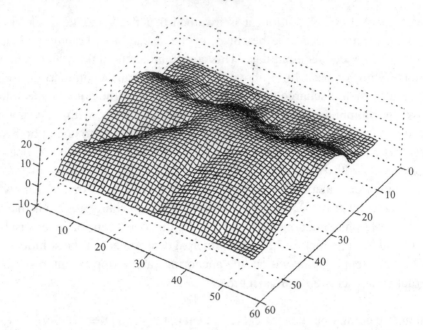

Figure 3.8. Calculated mean configuration of the water table in a 13.5-kilometer square region of Kent County, Md, located between the Chester River (lower boundary) and the Sassafras River (upper). The total curvature $\nabla^2 \zeta$ is generally negative, convex upward or dome-shaped, with slope discontinuities at the discharge lines along seasonal and semi-permanent streams.

region. The mean infiltration rate $\overline{W}$ was approximately 0.9×10^{-3} m/day. Values of the transmissivity in the region were estimated from measured differences in water table elevations along selected simple flow paths in the region and use of the relation (3.31) below. This gave estimates from 340 to 510 m²/day, with an average of 420.

The water table configuration in the region was calculated from (3.25) as a function of time. As shown in Figure 3.8, it is clearly anchored to the surface topography along the semi-permanent streams but its shape is characteristically quite different from that of the topography, with convex-upward, flattened domes between wedge-shaped valleys along the lines of discharge, as anticipated above. Field measurements of water table levels at a series of test wells in the region were generally within 1 m of those calculated, and this provides some assurance of the usefulness of the analysis.

The calculated flow paths and mean groundwater contours are illustrated in Figure 3.9, from which it is clear that the groundwater flow domains are of two distinct kinds. The boundaries are defined by tracing back the flow lines from the points of stream discharge into the estuary. One kind of domain originates broadly from infiltration near the groundwater divide where the water table slope

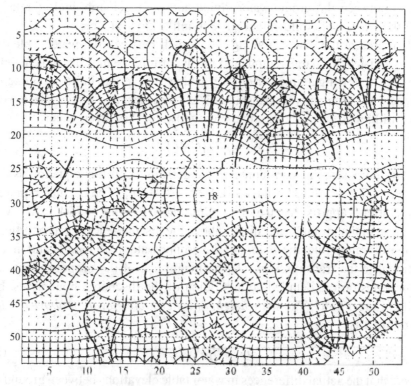

Figure 3.9. Contours of water table elevation as in Figure 3.8, with the pattern of sub-surface flow direction defining flow domains that either (i) originate near the groundwater divide and discharge into seasonal streams or (ii) receive their input from the slope regions and discharge directly by seepage through the shoreline of the estuary.

is small and the fluid moves relatively slowly. The flux is augmented by continued infiltration and accelerates (i.e. diverges longitudinally) along its path as it converges laterally toward discharge into one or another of the semi-permanent drainage streams. The other kind of domain avoids the streams, receives most of its input from the slope regions, and discharges directly along the shoreline into the estuary from springs or by seepage. They are exemplified by the uniformly *divergent* flows in the peninsulas along the Sassafras River to the north and the broad regions between the river catchment areas to the south. In this region, they apparently provide about 30% of the total groundwater flux to the estuary.

3.2.4 Scales of water table elevation; relaxation, emergence and recharge times

The structural simplicity of the governing equation (3.25) allows us to specify useful and robust scales, or general magnitudes, of the difference in water table elevations between the groundwater divide and the discharge, and of relaxation times of the water table following a pulse in infiltration rate, etc.

Let l_F represent the distance along a flow path between the water table divide and the associated discharge region, and h_W the difference in water table heights at the two ends following a time period over which the average infiltration rate is $\overline{W}$. In the mean, the time derivative term in (3.25) vanishes, and it becomes a balance between infiltration from above and the groundwater flux divergence:

$$\overline{W} = -\nabla\{C\nabla\zeta\}. \tag{3.29}$$

For scaling purposes, one can represent the derivatives in this equation as differences, so that $\nabla\zeta \sim -h_W/l_F$ and $\nabla \cdot \{C\nabla\zeta\} \sim Ch_W/l_F^2$. The balance represented by this equation is approximated by $\overline{W} - Ch_W/l_F^2 \sim 0$, or $h_W \sim l_F^2\overline{W}/C)$. In words, the elevation of the groundwater divide above the discharge level is proportional to the mean infiltration rate, the *square* of the groundwater path length, and inversely proportional to the transmissivity C. The magnitude of the proportionality coefficient depends on the detailed configuration of the discharge boundaries, but is generally of order unity. The *scale height of water table elevation above discharge* is then defined as

$$h_W = l_F^2\overline{W}/C. \tag{3.30}$$

We expect that the actual differences in water table elevations between groundwater divide and discharge will differ from this scale height only by a numerical factor of order unity, as exemplified by (3.26) and the corresponding calculation for circular geometry.

Although the causality involved is appropriately expressed in (3.30), i.e. a given mean infiltration rate, length of flow path and transmissivity produces a maximum water table elevation above discharge given by this expression, a rearranged form is probably of more practical use:

$$C = l_F^2\overline{W}/h_W. \tag{3.31}$$

In a field situation, it may be relatively easy to measure or estimate h_W to within 10 or 20% from surveyed well-water levels and l_F from the discharge boundary configuration. This formula then allows a very simple and relatively accurate method for determining the transmissivity C of the aquifer, avoiding the almost impossible task of estimating it from permeability profiles.

The *relaxation time* is the time the system takes to recover from a sudden perturbation. Time-dependent solutions to (3.25) (see, for example, Carslaw & Jaegar, 1980) are more complex than the simple steady solutions given above but the relaxation time scale can be found from the equation's structure. Suppose that a steady-state situation is interrupted by an impulsive infiltration event such as a drenching flood. The pulse of infiltration traverses the vadose zone relatively quickly, and raises the water table elevation essentially uniformly in the basin. In any diffusive (or conductive) system specified by an equation of the type (3.25), where a first-order time derivative is balanced by second-order space derivatives, the relaxation time is half the square of the length scale divided by the "diffusivity," here the combination C/ϕ of transmissivity and porosity. Thus

$$T_{RX} = \frac{\phi l_F^2}{2C}, \tag{3.32}$$

where l_F is the maximum diffusion distance, the distance from the groundwater divide to the discharge. A convenient and somewhat surprising formula expressing the relaxation time in terms of the water table scale height and the mean infiltration rate, both easily measured quantities, is found by combining (3.30) and 3.32):

$$T_{RX} \sim \frac{\phi h_W}{\overline{W}}, \tag{3.33}$$

with a constant of proportionality of about 1/2. This last expression has an interesting physical interpretation of the relaxation time as the time required for the mean infiltration to provide enough water to saturate the medium to the depth h_W.

In the aquifer system between the Chester and Sassafras Rivers the aquifer path lengths l_F are about 3 km and with the transmissivity $C \approx 420 \, \text{m}^2/\text{day}$, from (3.32) the relaxation time T_{RX} is about 8.8 years. This is the time scale for the groundwater level near the divide to return substantially towards its average configuration following a prolonged wet spell, but for points closer to the discharge, the appropriate length scale in (3.33) is the distance to the discharge, rather than l_F. At a distance of 1 km, the relaxation time is about 1 year. In either case, the time taken for the infiltration to traverse the vadose zone above the water table is considerably shorter, so that fluctuations in water level in un-pumped wells, particularly near the groundwater divide, are expected to be generally asymmetrical in time, with rapidly rising levels in wet spells and much more gradual declines in dry periods.

Another important time scale in aquifer flow is the *emergence time*, that is, the time it takes for water, entering the water table at a particular location, to travel to discharge. If a pulse of chemically passive but contaminated water is introduced across the water table at some location, it moves with the interstitial fluid toward the discharge at ever-increasing depths in the aquifer, because of the continuing infiltration from above. The time that it takes to travel along its streamline to the discharge boundary, the emergence time, is a function of the infiltration rate, the aquifer thickness and the fractional distance of the point of injection along the streamline from groundwater divide to discharge. One can see intuitively that the closer the point of injection is to the divide, the longer it will take to purge the contaminant from the aquifer.

Suppose that at some time t_0, a soluble, but chemically passive tracer is injected across the water table at a distance s_0 from the groundwater divide. In general, at any distance s from the divide, the upstream infiltration $s\overline{W}$ is balanced by the volume flux along the aquifer, ud, where u is the longitudinal transport velocity. The interstitial fluid velocity along the stream tube, ds/dt, is then $\phi^{-1}u = \phi^{-1}(\overline{W}/d)s$, so that

$$\frac{ds}{dt} \approx \left(\frac{\overline{W}}{\phi d}\right) s.$$

Since $s = s_0$ when $t = t_0$, after the time $(t - t_0)$, the tracer pulse has moved to a distance s from the groundwater divide given by

$$s(t) = s_0 \exp\left(\frac{\overline{W}}{\phi d}(t - t_0)\right). \tag{3.34}$$

Contaminated fluid emerges at the discharge, a distance $s = l_F$ from the ground-water divide, at what we call the emergence time T_E, say, given from the solution above as

$$T_E = \frac{\phi d}{\overline{W}} \ln\left(\frac{l_F}{s_0}\right). \tag{3.35}$$

The first factor $T_{RC} = \phi d/\overline{W}$, say, can be interpreted as the mean time it would take for the infiltration to supply a volume of water per unit surface area equal to the total pore volume per unit area throughout the depth of the aquifer. This, the *recharge time*, is a generally useful characteristic time scale in aquifer flows by setting not only the time scale for groundwater age and contaminant discharge, but it also appears naturally in a number of other Lagrangian aquifer characteristics discussed in the next section. The recharge times of aquifers in moderate climates are usually of the order one or two decades. For example, in an aquifer of porosity 0.3 and thickness 20 m, with a mean annual infiltration rate of 0.5 m/s, the recharge time is about 12 years. The second (logarithmic) factor is of course positive since the distance l_F from groundwater divide to discharge is always greater than the distance s_0 from the groundwater divide to the injection point. When the injection point is near the discharge, l_F/s_0 in (3.35) is only a little larger than unity, the numerical value of the logarithm is small and the emergence time is very short. When the injection is in the upper one-third (approximately) of the aquifer length, as was the case in measurements on Cape Cod by Garabedian *et al.* (1991), the emergence time is greater than the recharge time, $\phi d/\overline{W}$. If the injection occurs very near the groundwater divide, l_F/s_0 is large and the emergence time may be a substantial multiple of the recharge time, T_{RC}.

3.2.5 Groundwater age distribution in an aquifer

The "age" of a groundwater sample is defined as the time interval since it entered the groundwater and lost contact with the atmosphere. Dissolved chemical or radiological indicators will "follow the flow" with accuracy if dispersion or internal mixing of the interstitial fluids can be neglected. In particular, some CFCs (chloroflurocarbons) are chemically stable, manmade compounds that have been manufactured since about 1940, and their concentrations found in groundwater have been used as indicators of the time interval since the infiltrating water was recharged and isolated from the atmosphere. Dunkle *et al.* (1993) give an extensive discussion and evaluation of their use in the Delmarva Peninsula in the Atlantic Coastal Plain of the United States.

If the composition of a sandy aquifer is uniform except possibly for isolated much *less* permeable silt or clay lenses, the dispersion about the mean streamlines is small, and a very simple and again somewhat surprising relation can be found for the age of the groundwater in terms of the mean infiltration rate, the porosity, the aquifer thickness and the sampling depth. Remarkably, this relation is independent of the geography of the groundwater flow region but does assume the gentle relief, thin layer conditions, specifically that the aquifer thickness d from the water table to the basement varies only gradually along the flow path and is small compared with the longitudinal extent l_F of the groundwater flow. The horizontal velocity component is then independent of depth. The mean infiltration $\overline{W}$ is assumed uniform over the domain and so the vertical component of the mean interstitial velocity $\overline{v}_Z$ is equal to $-\overline{W}/\phi$ at the water table, decreases linearly with depth and vanishes at the aquifer basement $z = -d$:

$$\overline{v}_Z = -\frac{\overline{W}}{\phi}\left(1 + \frac{z}{d}\right). \tag{3.36}$$

Now, express this equation in Lagrangian form in which the position of the fluid element is $z(t)$ and its vertical velocity $\overline{v}_Z$ is dz/dt, so that it becomes

$$\frac{dz}{dt} = -\frac{\overline{W}}{\phi}\left(1 + \frac{z(t)}{d}\right). \tag{3.37}$$

The variables z and t are separable, and if time is measured from zero as the fluid element crosses the water table, the solution provides the groundwater age τ_A for fluid elements at the level $-z$ (below the water table), in terms of the recharge time scale and the fractional depth z/d:

$$\tau_A = \frac{\phi d}{W} \ln\left\{(1 + z/d)^{-1}\right\} = T_{RC} \ln\left\{(1 + z/d)^{-1}\right\}. \tag{3.38}$$

This is zero at $z = 0$ (at the water table, the interstitial water is always "new") and becomes indefinitely large at the basement, as $z \to -d$. The singularity at the bottom is artificial since no basement is perfectly smooth; the water age close to the bottom of the aquifer may be very large, but not infinite. Note that the basement depth d may vary gradually throughout the basin, but apart from that, the result (3.38) is independent of the geographic configuration of the aquifer. The reason for this somewhat counter-intuitive conclusion is that the thin layer geometry forces the transport velocity along the aquifer to be uniform in depth. No matter where they are, the older fluid elements continue to be displaced downward, driven by the continuing infiltration across the water table, until they are released upward in the discharge zone.

In Figure 3.10, the theoretical result (3.38) is compared with field measurements of CFC-model groundwater ages at various locations and depths in the recharge area of the Chesterville Branch, Kent County, Maryland, made by Reilly *et al.* (1994) and Böhlke and Denver (1995), and in Cape Cod by Böhlke, Smith and Miller (2006). Each aquifer region seems to be reasonably uniform with a depth from water table to basement of 20–22 m in Maryland and about 40 m in Cape Cod. The recharge times were about 17.5 and 25 years, respectively. Although there is a good deal of scatter, as is characteristic of field measurements, the points from the two sets of measurements fall into a reasonably consistent cloud about the continuous curve representing the theoretical result (3.38). Some shallow, anomalously old samples may have been taken in a discharge region or in a relatively impermeable lens where the interstitial water is almost immobile, and deep, young samples may have been associated with sloping, more permeable conduits that attract the flow from the surroundings. The scaling uses measured values of the mean infiltration rates and aquifer thicknesses, and, with *no fitting parameters needed*, the agreement with the field measurements is surprisingly good.

3.3 Dispersion and transport of marked fluid

3.3.1 Measurements of permeability variations in sandy aquifers

As early as 1975, Freeze noted that in many field formations, the natural logarithm of the permeability (referred to a convenient reference value) appeared to

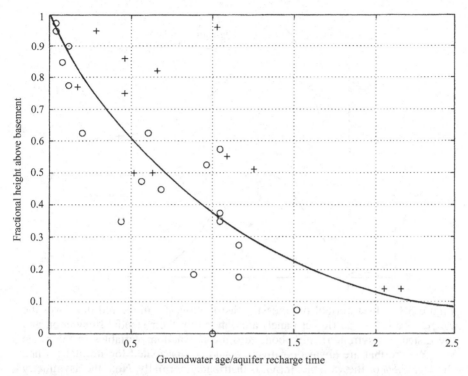

Figure 3.10. The distribution with depth of CFC water ages, referred to the appropriate recharge time, as measured by Böhlke and Denver (1995) on the Maryland Eastern Shore (+) and by Böhlke, Smith and Miller (2006) in Upper (i.e. western) Cape Cod (o). The curve represents the theoretical result (3.38). The recharge time $T_{\mathrm{RC}} = \phi d/\overline{W}$, was estimated from independent measurements on each aquifer; no curve fitting is involved.

be distributed in an approximately Gaussian manner. Subsequently, extensive and important measurements by Hess, Wolf and Celia (1992) have established the characteristic magnitude and geometrical structure of the hydraulic conductivity variations in a sand and gravel aquifer on Cape Cod, Massachusetts, near Otis Air Force Base. Their site was chosen because of the existence of contaminant plumes where gasoline appears to have leaked into the groundwater during World War II. Nearly 1500 permeability measurements were made in a carefully sited grid of wells by both borehole flow-meter tests and permeameter analysis of extracted cores. Some systematic differences were found between the two sets of measurements, the values of hydraulic conductivity K found from the borehole flow-meter tests being generally consistent with previous results from this aquifer, while those found from measurements on the cores were consistently smaller. Hess *et al.* suggest that the core samples may have been compacted during the coring and retrieval processes, so that the two sets of data were analysed separately. The local

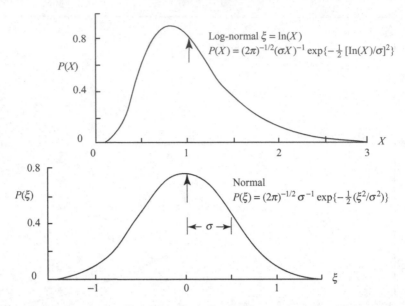

Figure 3.11. In a normal or Gaussian distribution, the measured data over the range $-\infty < \xi < \infty$ (lower panel) are defined by their standard deviation and are distributed symmetrically about zero mean. Random variables in the range $0 < X < \infty$ that are always positive are said be distributed log-normally when the *logarithm* of the variable, $\ln(X)$, is distributed normally. Note the asymmetry of the log-normal distribution (upper panel) with the most frequent occurrences being smaller than the mean and a few being much larger than the mean.

hydraulic conductivity measurements from the flow meters had a geometric mean of 0.11 cm/s and those from the permeameter cores 0.035 cm/s.

The separate distributions were found to be closely log-normal at the 95% confidence level. In the log-normal distribution of a variable that is always positive but otherwise random (such as a permeability field), it is the logarithm of the variable that is distributed normally. Log-normal and normal (or Gaussian) distributions with standard deviation σ are illustrated in Figure 3.11, with the two median values aligned vertically so that on either side there is an equal probability of occurrence. In the log-normal distribution, the range of values less than the median is between 0 and 1 and the probability density is highest. High values of the variable continue to occur over a wider range than in a normal distribution, although with increasingly lesser frequency. The distribution is then characterized by the occurrence of a few very large values amid a much larger number of relatively small values. The observation that in a natural aquifer, the permeability distribution is closely log-normal, implies that a few highly permeable regions or conduits are characteristically distributed in a variable, but generally less permeable matrix, with possibly a few distributed, almost impermeable nodules.

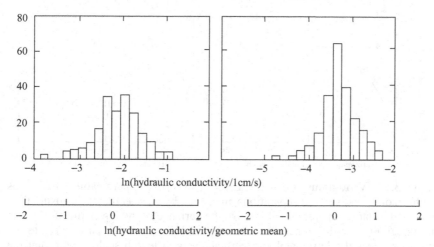

Figure 3.12. Histograms of subsets of the natural logarithms of hydraulic conductivity relative to 1 cm/s (upper scale), and to the geometric mean of the subset (lower scale), measured by Hess *et al.* (1992) in a sand and gravel aquifer in Cape Cod, Massachusetts. The left-hand panel shows the distribution from 194 flow meter measurements and the right-hand panel shows results from 213 permeameter measurements.

The log-permeability $l(x)$ is found to be a dimensionless field variable

$$l(x) = \ln\left(K(x)/K_0\right) = \ln\left(k(x)/k_0\right), \tag{3.39}$$

in which the reference value of the hydraulic conductivity K_0 was chosen arbitrarily by Hess *et al.* to be 1 cm/s. Their measured distributions of $l(x)$ from the two sets of data are shown in Figure 3.12, and both are reasonably Gaussian. The variance of $l(x)$ from the flow-meter data was found to be 0.24 and that from the cores was 0.14. Note, incidentally, that the natural logarithm of the permeability fluctuations is a field quantity that occurs in the basic relations (3.9) and (3.11) connecting the velocity and pressure fields separately to the permeability field, so that log-normality of the permeability is consistent with normal distributions of these other fields.

Hess *et al.* expressed the spatial structure of the variations in hydraulic conductivity in terms of the "variogram" $V(r)$ (in statistical hydrodynamics usually called the structure function, c.f. Monin and Yaglom, 1975) of the spatial distribution $l(x)$. This is defined as half of the mean square *difference* in $l(x)$ between two points x and $x + r$ in the medium, separated by the vector distance r:

$$V(r) = \tfrac{1}{2}\overline{[l(x + r) - l(x)]^2}, \tag{3.40}$$

where the overbar denotes averaging. The structure function is a useful measure of meso-scale variability, being relatively insensitive to larger (aquifer-scale)

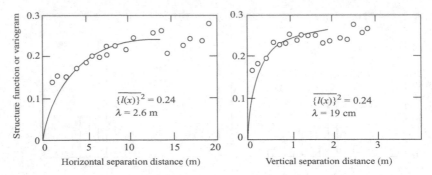

Figure 3.13. Variograms or structure functions $V(r)$ of log-permeability variations for horizontal and vertical separations, measured by Hess *et al.* (1992) using flow meters. The large separation values give the variance of the log-permeability and the integral of the differences between this and the measurements closer to the origin provides the horizontal and vertical integral length scales. Note that the integral length scale of the log-permeability variations in the horizontal direction is about an order of magnitude larger than in the vertical.

variations in local mean properties. Analysis of the measurements in terms of this function provides two of the most significant measures of the randomness, the mean square magnitude of the variations and their characteristic vertical and horizontal integral length scales. It can be expressed in terms of the spatial correlation function $f_{Lk}(r)$ of the variations in log-permeability, which is the ratio of the covariance of log-permeability, a function of the spatial separation r, to the variance:

$$f_{Lk}(r) = \overline{\{l(x)l(x+r)\}}/\overline{\{l(x)\}^2}.$$

This is dimensionless, has the value 1 when $r = 0$ and is expected to decrease to zero as the separation increases beyond the characteristic length scale of these variations. From (3.40), the structure function or variogram assumes the form

$$V(r) = \overline{\{l(x)\}^2}[1 - f_{Lk}(r)]. \qquad (3.41)$$

Since $f_K = 1$ when $r = 0$, from this definition, $V(0) = 0$. Note, however, that field measurements of local permeability differences between closely spaced points are inherently inaccurate because of interference effects, as is evident from the first few points of Figure 3.13. The magnitude of the mean-square log-permeability variations on a local scale can be found from the value of $V(r)$ at large separations (when f is small) and the correlation length can be estimated from the distance over which $V(r)$ decreases towards zero as the origin is approached.

Spatial horizontal correlation scales λ were estimated by fitting an assumed exponential function $1 - \exp(-r/\lambda)$ to the discrete measurement points along each transect, and best fit values ranged from about 2.6 m from flowmeter data to 2.0 m from permeameter data. In some measurements of the horizontal structure functions or variograms, $V(r)$ decreased at small separations but not to zero, possibly because of unresolved small-scale variability (one meter or less). There was little evidence of horizontal anisotropy. Vertical correlation scales, 0.18 m, were smaller by a factor of about 15. Although these numerical values are specific to this site, their dominant characteristics, with approximately equal horizontal correlation length scales and a much smaller vertical correlation length, are expected to be generally representative of other sandy aquifers as well. The results of these remarkable pioneering measurements probably remain the best available characterization of small-scale permeability variations in aquifers of this kind.

3.3.2 *Measured dispersion of injected tracers over sub-kilometer scales*

Spatial variations in the local mean interstitial velocity field in an aquifer containing random variations in log-permeability have not been measured directly, but a number of flow and dispersion measurements have been made by injecting passive tracers that move with the interstitial fluid velocity and following their movements using arrays of sampling sites. The durations over which measurements continued (from 1 to 3 years) were small compared with the aquifer recharge times and the path lengths involved were relatively short, 200–300 m, much smaller than the total aquifer flow path lengths. The water table slope in each of the study areas was very nearly uniform and the mean interstitial flow velocity approximately horizontal. Among the earliest of these measurements were those made in an unconfined surface aquifer close to a landfill near Borden, Ontario, by Mackay *et al.* (1986), by Freyberg (1986) and by Roberts, Goltz and Mackay (1986). This aquifer was described as relatively homogeneous, with clean, well-sorted fine- to medium-grained sand of hydraulic conductivity about $(5 - 10) \times 10^{-3}$ cm/s, with generally horizontal bedding and some thin horizontal layers or lenses, less permeable by a factor of about 10. A dense three-dimensional array of over 5000 sampling points was installed to track a pulse injection of $12\,m^3$ of conservative inorganic and organic solvents over a period of about 3 years. Freyberg noted that the trajectory of the marked fluid through the aquifer was slightly curvilinear, concave upward, as is consistent with the preceding discussion and the streamline shapes of Figure 3.5 that characterize continuous, distributed infiltration. At this site, the dispersion was markedly non-Gaussian (non-Fickian) and in some places bi-modal in the vertical, presumably because of the layering. Mackay *et al.* found that the spreading in the horizontal longitudinal direction was *much more pronounced* than in the horizontal transverse direction and that vertical spreading was very much smaller again. This finding of extreme anisotropy in the marked fluid region has been confirmed by many others, and indeed has turned out to be a ubiquitous property of aquifer dispersion, despite the approximate isotropy in the structure of horizontal variations in aquifer permeability noted above.

Further field measurements have been made by LeBlanc *et al.* (1991), by Garabedian *et al.* (1991) near Otis Air Force base on Cape Cod, Massachusetts, and by Jensen, Bitsch and Bjerg (1993) in a sandy aquifer in Denmark. These have provided more quantitative estimates of the dispersion characteristics of various tracers in media of this kind. In these measurements, after an initial interval the dispersing clouds generally spread in time as $t^{1/2}$, as in classical (Fickian) diffusion, and in accordance with (2.75). The apparent dispersivity was quite anisotropic, Garabedian *et al.* reporting average values of 0.96 m in the longitudinal flow direction, 1.8 cm in the transverse horizontal direction and 1.5 cm in the vertical, both smaller than the longitudinal dispersivity by a factor of about 50. Corresponding measurements from Hess *et al.* were 0.52 m and 0.018 m, again differing by a large factor.

In the measurements reported by Leblanc *et al.* (1991), a non-reactive bromide tracer cloud was injected just below the water table at a site approximately 10 km from the groundwater divide. It was followed for 461 days, during which time the centroid of the cloud had moved about 200 m horizontally at the rate of 0.42 m/day, very close to the interstitial fluid velocity calculated from the water table slope, the measured hydraulic conductivity (110 m/day) and the porosity (0.39). As shown in Figure 3.14, redrawn from their paper, the bromide cloud had spread over a distance of more than 80 m in the direction of flow after 461 days, the end of the measurement period. The width of the cloud was then about 14 m and the thickness only about 4–6 m. The cloud had also drifted downward about 2.3 m, as a result of rainwater infiltration and possibly also a slight excess density in the bromide solution near the injection point.

These field measurements pose the following question. While the spatial distribution of log-permeability is almost isotropic in horizontal planes, why is the longitudinal dispersion in flow through an aquifer so much larger than that in transverse directions? This is clearly a robust phenomenon, whose salient characteristics should be comprehensible in terms of Darcy flow characteristics in a medium of variable permeability and the overall minimum dissipation constraint. The variance σ^2 and the vertical and horizontal integral scales are the two most important statistics of the log-permeability variations that influence the spreading of dissolved solutes, but despite the considerable effort devoted to this problem by Gelhar and Axness (1983), Gelhar (1986), Dagan (1988) and others, the detailed connection does not seem to be widely understood.

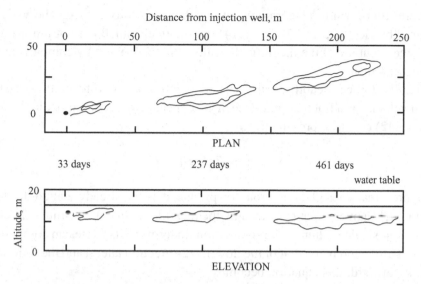

Figure 3.14. Downstream advection and spreading of an injected pulse of a passive bromide tracer from the point source marked on the left, which is approximately 10 km south of the groundwater divide in south-western Cape Cod. In the region shown, the mean interstitial fluid velocity was almost uniform over the 250 m range. The vertical scale is exaggerated by a factor 2. Longitudinal dispersion is very much greater than lateral dispersion, both horizontal and vertical. The measurements were made by Leblanc *et al.* (1991).

3.3.3 *Flow through a spatially random permeability field*

Spatial variations in the transport velocity field are directly associated with the mean velocity and the variations in the log-permeability field through the rotation vector expression (3.9), according to which a random, rotational and presumably dispersive flow is induced when streamlines cut across local random gradients of permeability. It is convenient in theoretical analyses to refer the permeability distribution $k(x)$ to its geometric mean, k_0, say. The log-permeability field $l(x) = \ln\{k(x)/k_0\}$ is then a stationary random function of position *with zero mean*, as shown in the lower scale of Figure 3.12. The mean flow, U, passing through this random permeability field, generates random, rotational patterns of transport velocity that satisfy (3.9) which, for convenience is reproduced here:

$$\nabla \times u = \nabla l(x) \times (U + u). \tag{3.42}$$

Solving this equation is difficult and awkward to do in physical space because of its structure, but when the variance of the log-permeability $l(x)$ is fairly small, the problem is much simpler in Fourier space since the differential equations become linear and transform to algebraic equations that are easily solved. The variances measured by Hess *et al.* (0.24 and 0.14) *are* relatively small but not infinitesimal, so that the accuracy of the theoretical results will be assessed by comparison with field measurements.

When the log-permeability variances are small (compared to 1), the variations in local velocity as a fraction of the mean are expected to be correspondingly small, so that (3.42) can be approximated by

$$\nabla \times \boldsymbol{u} = \nabla \left\{ l(\boldsymbol{x}) \right\} \times \boldsymbol{U}. \tag{3.43}$$

This approximation neglects the rotation produced by the already scattered velocity variation $\boldsymbol{u}$, relative to that produced by the mean $\boldsymbol{U}$, so that it constitutes in effect, a first-order, single scattering approximation. In terms of its Cartesian components, with the 1-direction being that of the flow, the 2-direction lateral and the 3-direction vertically upward, this equation becomes

$$(\nabla \times \boldsymbol{u})_1 = \frac{\partial u_3}{\partial x_2} - \frac{\partial u_2}{\partial x_3} = 0$$

$$(\nabla \times \boldsymbol{u})_2 = \frac{\partial u_1}{\partial x_3} - \frac{\partial u_3}{\partial x_1} = U \frac{\partial l(\boldsymbol{x})}{\partial x_3}$$

$$(\nabla \times \boldsymbol{u})_3 = \frac{\partial u_2}{\partial x_1} - \frac{\partial u_1}{\partial x_2} = -U \frac{\partial l(\boldsymbol{x})}{\partial x_2}. \tag{3.44}$$

The incompressibility condition is $\nabla \cdot \mathbf{u} = 0$.

The transport velocity variation and the log-permeability field are random functions with zero mean, with Fourier–Stieltjes transforms defined as

$$u_i(\boldsymbol{x}) = \int_{\kappa} \exp[i\boldsymbol{\kappa} \cdot \boldsymbol{x}] dZ_i(\boldsymbol{\kappa}), \, i = 1, 2, 3, \tag{3.45}$$

$$l(\boldsymbol{x}) = \ln \left\{ k(\boldsymbol{x})/k_0 \right\} = \int_{\kappa} \exp[i\boldsymbol{\kappa} \cdot \boldsymbol{x}] d\lambda(\boldsymbol{\kappa}). \tag{3.46}$$

The Fourier–Stieltjes transforms $dZ_i(\boldsymbol{\kappa})$ and $d\lambda(\boldsymbol{\kappa})$ are interpreted as the sums of contributions to $u_i(\boldsymbol{x})$ and $l(\boldsymbol{x})$ from an element of volume $d\kappa$ near the point $\boldsymbol{\kappa}$ in wave-number space, and the integral (or sum, or superposition) is over all wave-number space. They are a particular variety of Fourier transform for random functions whose basic mathematics was given by Kampé de Férier (1939) with applications to analysis of random noise by Rice (1944) and to turbulence by Batchelor (1953) and others. A detailed discussion is beyond the scope of this book, though a few considerations may provide some plausibility. The transform itself, $d\lambda(\kappa)$, is random and not differentiable with respect to κ – it resembles a dense forest of delta-function spikes – but the definitions above can be differentiated with respect to the spatial variables. Thus, in the first of the set (3.44),

$$\frac{\partial u_3}{\partial x_2} = i \int_{\kappa} \exp(i\kappa \cdot x)\kappa_2 dZ_3(\kappa), \qquad \text{etc.,}$$

so that this equation becomes

$$\frac{\partial u_3}{\partial x_2} - \frac{\partial u_2}{\partial x_3} = i \int_{\kappa} \exp(i\kappa \cdot x)\left[\kappa_2 dZ_3 - \kappa_3 dZ_2\right] = 0.$$

The Fourier transform of zero is zero, and consequently

$$\kappa_2 dZ_3(\kappa) - \kappa_3 dZ_2(\kappa) = 0. \tag{3.47a}$$

Likewise,

$$\kappa_3 dZ_1(\kappa) - \kappa_1 dZ_3(\kappa) = U\kappa_3 d\lambda(\kappa)$$
$$\kappa_1 dZ_2(\kappa) - \kappa_2 dZ_1(\kappa) = -U\kappa_2 d\lambda(\kappa) \tag{3.47b}$$

where $d\lambda(\kappa)$ is the Fourier–Stieltjes transform (3.46) of the log-permeability field. The simultaneous differential equations (3.44) have thus been transformed to the set of simultaneous algebraic equations (3.47a,b) and the incompressibility condition $\nabla \cdot u = 0$ similarly transforms to

$$\kappa_1 dZ_1 + \kappa_2 dZ_2 + \kappa_3 dZ_3 = 0. \tag{3.48}$$

Algebraic manipulation (i.e., multiply the first of (3.47b) by κ_3, the second by κ_2 and subtract, then substitute from (3.48)) gives

$$\left(\kappa_1^2 + \kappa_2^2 + \kappa_3^2\right) dZ_1(\kappa) = \left(\kappa_2^2 + \kappa_3^2\right) U d\lambda(\kappa),$$

so that

$$dZ_1(\kappa) = \left\{1 - \frac{\kappa_1^2}{\kappa^2}\right\} U d\lambda(\kappa), \tag{3.49a}$$

where $\kappa^2 = \kappa_1^2 + \kappa_2^2 + \kappa_3^2$. Substitution back into (3.47b) provides corresponding expressions for the other two components,

$$dZ_2(\kappa) = -\frac{\kappa_1\kappa_2}{\kappa^2} U d\lambda(\kappa) \qquad \text{and} \qquad dZ_3(\kappa) = -\frac{\kappa_1\kappa_3}{\kappa^2} U d\lambda(\kappa) \tag{3.49b}$$

These expressions give the Fourier–Stieltjes transforms of the three components of the velocity field in terms of the corresponding transform of the log-permeability field that generated them. They are building blocks for the construction of wave-number spectra, covariances and variances of the velocity distribution in terms of the log-permeability field. To do this, multiply (3.46) by its own complex conjugate taken at the point $\mathbf{x} - \mathbf{r}$, and then average the product to form the covariance of the log-permeability field:

$$L(r) \equiv \overline{l(x)l(x+r)} = \int\limits_{\kappa} \int\limits_{\kappa'} \exp[i(\kappa' - \kappa) \cdot x] \exp i(\kappa' \cdot r)\overrightarrow{d\lambda^*(\kappa)d\lambda(\kappa')}.$$

Now, we assume that the log-permeability field is spatially homogeneous, so that the double integral on the right-hand side must be independent of x. This can occur if the Fourier–Stieltjes components at *different* wave-numbers κ and κ' are uncorrelated, so that the only contributions to the double integral are from points along the trajectory $\kappa = \kappa'$. For these, the first exponential factor is just one, and the last equation becomes

$$L(r) = \int \exp(i\kappa \cdot r)\overline{d\lambda(\kappa)d\lambda^*(\kappa)} = \int \Lambda(\kappa)\exp(i\kappa \cdot r)d\kappa_1 d\kappa_2 d\kappa_3, \quad (3.50)$$

where the log-permeability spectrum,

$$\Lambda(\kappa) = \frac{\overline{d\lambda(\kappa)d\lambda^*(\kappa)}}{d\kappa_1 d\kappa_2 d\kappa_3} \quad (3.51)$$

is the Fourier transform of the log-permeability covariance, with physical dimensions of L^3. This function is called a spectrum because it specifies the distribution in wave-number space of contributions to the total mean-square log-permeability, just as a sound spectrum specifies the distribution in frequency of contributions to total acoustic energy. From (3.50) with $r = 0$ and writing the element of volume in wave-number space $d\kappa_1 d\kappa_2 d\kappa_3$ as $d\kappa$, we have

$$L(0) = \overline{l(x)^2} = \int \Lambda(\kappa)d\kappa; \quad (3.52)$$

the mean square (or variance) of the log-permeability distribution is the integral of the spectrum over the entire three-dimensional wave-number space. The inverse transform associated with (3.45) gives the wave-number spectrum in terms of log-permeability covariance in the physical domain:

$$\Lambda(\kappa) = (2\pi)^{-3} \int L(r)\exp(-i\kappa \cdot r)dr. \quad (3.53)$$

Note the normalizing factor and negative imaginary exponent associated with the inverse transform.

Similarly, the covariance tensor of the velocity field is

$$R_{ij}(r) = \overline{u_i(x)u_j(x+r\}} = \int \Phi_{ij}(\kappa)\exp(i\kappa \cdot r)d\kappa \quad (3.54)$$

where, in terms of the Fourier–Stieltjes transform (3.45), the spectrum tensor is

$$
\begin{aligned}
\Phi_{ij}(\kappa) &= \frac{\overline{dZ_i(\kappa)dZ_j^*(\kappa)}}{d\kappa_1 d\kappa_2 d\kappa_3} \\
&= (2\pi)^{-3} \int R_{ij}(r) \exp(-i\kappa \cdot r) dr.
\end{aligned}
\tag{3.55}
$$

Thus $\Phi_{ij}(\kappa)$ is also the inverse transform of the velocity covariance tensor.

The measurements of Hess *et al.* (1992) described earlier have established the primary statistical structure of the three-dimensional log-permeability field in the sand and gravel aquifer on Cape Cod, expressed by the covariance $L(r) = \overline{l(x)l(x+r)}$, whose Fourier transform, $\Lambda(\kappa)$, can be found from (3.53) and the measurements of the variogram or structure function made by Hess *et al.* The analyses above give the velocity spectrum tensor in terms of $\Lambda(\kappa)$, from which we can find the mean square velocity components and integral length scales in any direction that pertain to the dispersion measurements of Leblanc *et al.* (1991) and Garabedian *et al.* (1991). The most significant components of the velocity spectral tensor are those of flow in the longitudinal and transverse directions. From (3.55) and the solutions (3.49a), they are given by

$$
\begin{aligned}
\Phi_{11}(\kappa) &= U^2 \left\{ 1 - (\kappa_1/\kappa)^2 \right\}^2 \Lambda(\kappa), \\
\Phi_{22}(\kappa) &= U^2 \left\{ \kappa_1^2 \kappa_2^2 / \kappa^4 \right\} \Lambda(\kappa), \\
\Phi_{33}(\kappa) &= U^2 \left\{ \kappa_1^2 \kappa_3^2 / \kappa^4 \right\} \Lambda(\kappa),
\end{aligned}
\tag{3.56}
$$

where $\Lambda(\kappa)$ is the wave-number spectrum of the log-permeability distribution (3.53).

(i) Velocity variances

Mean-square values of the random velocity field induced by the log-permeability variations are found by integration of the spectra in (3.56) over all wave-numbers. Note that the second factor on the right-hand side of each equation involves the wave-number direction, not magnitude (in the first equation, for example, it is $(1 - \cos^2 \theta) = \sin^4 \theta$, where θ is the angle between the wave-number direction and the flow). Consequently, the velocity variances in each direction are proportional to the *square of the mean velocity* times the *variance of the log-permeability*, with numerical constants of proportionality that depend on these directional factors as well as on the degree of anisotropy of the permeability, but appear to be generally of order one. For example, if the log-permeability is isotropic, the first equation in (3.56) gives the mean square velocity variation in the flow or 1-direction. With the use of polar coordinates, the integration is

$$\overline{u_1^2} = \int_0^\infty \Phi_{11}(k)k^2 dk \int_0^\pi \sin\theta\, d\theta \int_0^{2\pi} d\phi = U^2 \int_0^\infty \Lambda(k)k^2 dk \int_0^\pi \sin^5\theta\, d\theta \int_0^{2\pi} d\phi$$

whence

$$\frac{\overline{u_1^2}}{U^2} = \frac{15}{8}\overline{l(x)^2} \tag{3.57}$$

the variance of the longitudinal velocity compared with the mean is almost twice the log-permeability variance! In contrast, from a similar calculation on the second or third of (3.56) gives

$$\frac{\overline{u_2^2}}{U^2} = \frac{\overline{u_3^2}}{U^2} = \frac{1}{2}\overline{l(x)^2}. \tag{3.58}$$

The corresponding variance of the transverse velocity referred to the mean is only half of the log-permeability variance (from Hess *et al.*, 1992, about 0.24). The relative smallness of the transverse velocity components is one factor that tends to reduce the lateral spreading of patches of chemically passive dissolved contamination, but not the most important one.

(ii) Dispersivities

The dispersivity coefficients (2.80) are expressed in the kinematic dispersion analysis as products of spatial velocity correlation scales in the flow direction and the ratios of mean square random flow velocities to mean velocity. In more basic terms, however, they are properties of the permeability structure of the medium. The connections can be found with the use of purely formal relations given by Batchelor (1953) connecting certain partial covariances in physical space and integrated spectra in Fourier space, together with our solutions (3.56) for the random flow field in terms of the permeability variations.

Take the inverse transform of (3.54) with respect to the variable r_1 only, and then take $i = j = 1$, as in Figure 3.15:

$$(2\pi)^{-1} \int_{-\infty}^\infty R_{11}(r_1, r_2, r_3)\exp(-i\kappa_1' r_1)dr_1$$

$$= \int\int \Phi_{11}(\kappa_1, \kappa_2, \kappa_3)\exp i(\kappa_2 r_2 + \kappa_3 r_3)d\kappa_2 d\kappa_3 \int_{-\infty}^\infty \exp\{i(\kappa_1 - \kappa_1')r_1\}dr_1$$

$$= \int_{-\infty}^\infty\int_{-\infty}^\infty \Phi_{11}(\kappa_1', \kappa_2, \kappa_3)\exp i(\kappa_2 r_2 + \kappa_3 r_3)d\kappa_2 d\kappa_3,$$

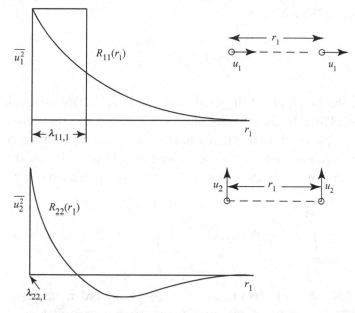

Figure 3.15. Longitudinal and lateral velocity correlations $R_{11}(r_1)$ and $R_{22}(r_1)$ between pairs of points in the direction of mean flow. The corresponding integral length scales $\lambda_{11,1}$ and $\lambda_{22,1}$ define the base of a rectangle of area equal to that under the curve; in the lateral correlation, this scale is very small or zero since the positive and negative contributions balance.

since the second integral in the previous line is a form of the Dirac delta function that replaces κ_1 by κ_1'. Now, assign the free variables $r_2 = r_3 = 0$ and $\kappa_1' = 0$, and we have:

$$\overline{u_1^2}\lambda_{11-1} = \int_0^\infty R_{11}(r_1, 0, 0)dr_1 = \tfrac{1}{2}\int_{-\infty}^\infty R_{11}(r_1, 0, 0)dr_1$$

$$= \pi \int \int \Phi_{11}(0, \kappa_2, \kappa_3)d\kappa_2 d\kappa_3 \qquad (3.59)$$

where, as illustrated in Figure 3.15, λ_{11-1} is the integral length scale in the 1-direction of the velocity variations in this same direction. Similarly, for the log-permeability field $l(x)$,

$$\overline{l(\mathbf{x})^2}\lambda_1 = \int_0^\infty L(r_1, 0, 0)dr_1 = \pi \int \int \Lambda(0, \kappa_2, \kappa_3)d\kappa_2 d\kappa_3, \qquad (3.60)$$

where λ_1 is the corresponding integral length scale.

The dynamical connection between these two is expressed by equations (3.56). When $\kappa_1 = 0$, the first of (3.56) reduces to

$$\Phi_{11}(0, \kappa_2, \kappa_3) = U^2\Lambda(0, \kappa_2, \kappa_3),$$

so that, from (3.59) and (3.60),

$$\alpha_{D-1} = \left(\overline{u_1^2}/U^2\right)\lambda_{11-1} = \overline{l(x)^2}\lambda_1. \tag{3.60a}$$

In words, the longitudinal dispersivity is equal to the mean-square value of the log-permeability times its integral length scale in the flow direction. This surprisingly simple result is apparently new and involves no arbitrary coefficients.

A similar calculation of the transverse dispersivity provides another surprise. For velocity components in the lateral 2-direction, we have as in (3.59),

$$\overline{u_2^2}\lambda_{22-1} = \int\limits_0^\infty R_{22}(r_1, 0, 0)dr_1 = \pi \int\int \Phi_{22}(0, \kappa_2, \kappa_3)d\kappa_2 d\kappa_3 = 0, \tag{3.61}$$

since from (3.56), $\Phi_{22}(\kappa) = 0$ when $\kappa_1 = 0$. To the first order, the lateral dispersivity vanishes! The correlation function $R_{22}(r_1, 0, 0)$ is equal to $\overline{u^2}$ when $r_1 = 0$ so that the vanishing of its integral requires that the function be negative over a significant range of r_1 as illustrated in Figure 3.15. This result is consistent with the overall minimum dissipation constraint. With the mean velocity in the 1-direction, the transverse velocity variations tend to reverse as the separation increases, reducing the lateral flow-path excursions that would add to the overall dissipation. The first-order theory based on equation (3.43) neglects the nonlinear term in equation (3.45) and becomes inaccurate when the mean-square variations in log-permeability are large. Nevertheless, the solutions remain qualitatively correct, as seen below; the lateral dispersivities are not precisely zero as predicted by equation (3.61), but are found to be one or two orders of magnitude smaller than the longitudinal values.

(iii) Comparison with field measurements

The extensive investigation in Cape Cod by Hess *et al.* (1992), described earlier, provided nearly 1500 individual measurements of hydraulic conductivity from which the mean intensity and scale of the log-permeability field were found. The flow meter tests seem to have been more reliable than the permeameter tests on extracted cores and gave a mean log-permeability variance of 0.24 and horizontal integral length scale of 3.5 m. With these measured values, the theoretical expression (3.60) predicts a longitudinal dispersivity of 0.84 m with no adjustable coefficients and transverse dispersivities of zero in both the vertical and lateral directions. Tracer measurements by Garabedian *et al.* (1991) found a mean longitudinal dispersivity of 0.96 m, very close to the theoretical value, and transverse values of 0.018 m horizontally and 0.0015 m vertically, both close to zero, as the theory predicts.

3.4 Layered media

3.4.1 Anisotropy produced by fine-scale layering

Consider a multi-layered permeable medium in which the local permeability varies in the vertical direction z but on a scale smaller than those we care to resolve. Averages over a resolution volume will include a sufficient number of layers that the overall effective permeability is expected to be anisotropic. Layered structures like this offer high permeability conduits to Horizontal flow and obstacles to any vertical flow that may be forced to traverse the retarding layers of low permeability. Working on a relatively large scale, Sternlof *et al.* (2006) used aerial mapping to identify patterns of dense, low-permeability compaction bands in outcrops of the Aztec Sandstone in the Valley of Fire State Park, 60 km north-east of Las Vegas. The highly laminated compaction bands, exposed over areas exceeding $160\,000\,\mathrm{m}^2$, have widths of order centimeters and individual lengths of order meters in which the saturated permeability is reduced by several orders of magnitude.

The usual model (see, for example, Bear, 1979, or Barenblatt, Entov and Ryzhik, 1990) supposes that the permeability in the layers varies *only* in the vertical direction, that the individual high- and low-permeability layers are *uniform and indefinitely long*. We assume also that the averaging scale, though large compared with the scale of variations in $k(z)$, is still much smaller than the formation thickness, and consider separately the cases of horizontal and vertical flows. For horizontal flow,

$$\mathbf{u}(z) = -\mu^{-1} k(z) \nabla_\mathrm{H} p,$$

so that within an individual averaging volume,

$$\frac{\mathbf{u}(z)}{k(z)} = -\mu^{-1} \nabla_\mathrm{H} p, \tag{3.62}$$

where the horizontal pressure gradient may vary on the formation scale, i.e., on scales l horizontally and h vertically, but is essentially constant within the averaging volume. In particular, in flow along the layering, the pressure gradient is assumed to be independent of z. Within the fine-scale layers, the local permeability and horizontal velocity vary proportionately and the flow is predominantly in the high permeability layers. The mean horizontal transport velocity averaged over the resolution height interval h_R is

$$\bar{\mathbf{u}}(z) = h_\mathrm{R}^{-1} \int_0^{h_\mathrm{R}} \mathbf{u}(z) dz = -\mu^{-1} (\nabla_\mathrm{H} p) \left(h_\mathrm{R}^{-1} \int_0^{h_\mathrm{R}} k(z) dz \right)$$

$$= \mu^{-1} \bar{k}_\mathrm{H} \nabla_\mathrm{H} p, \qquad \text{say,} \tag{3.63}$$

where the effective permeability for the horizontal flow averaged over the resolution volume is simply the vertically averaged permeability in the interval h_R:

$$\bar{k}_H = h_R^{-1} \int_0^{h_R} k(z)dz. \tag{3.64}$$

In a binary medium with sub-intervals of high permeability k_{AQ} alternating with "retarding layers" of low permeability $k_{RL} \ll k_{AQ}$ that occupy the fraction β of the region, the resolution scale permeability

$$\bar{k}_H \approx (1 - \beta)k_{AQ}. \tag{3.65}$$

This is independent of k_{RL}, the permeability of the retarding layers, because little fluid moves along them.

The expression (3.64) and the approximation (3.65) giving the average permeability to flow *along* the layers, are expected to be quite robust. They do depend on the assumption that at least most of the aquifer layers are continuous through the averaging region, but only weakly upon the continuity of the retarding layers since, again, almost all of the flow moves around them.

Purely transverse flow *across* the layering is a different matter. The crucial assumption in the conventional model is that the retarding layers are indefinitely long, uniform and continuous, though this may be geologically unrealistic. Under this assumption, the channeling effect disappears and the vertical velocity within the averaging volume is assumed independent of z. The vertical pressure gradient needed to drive the flow now varies rapidly in z as $k(z)$ does.

$$\frac{\partial p}{\partial z} = -\frac{\mu w}{k(z)},$$

and the mean vertical pressure gradient, averaged over the interval h_R is

$$\frac{\partial \bar{p}}{\partial z} = -\mu w \left\{ \frac{1}{h_R} \int_0^{h_R} (k(z))^{-1} dz \right\} = -\mu w / \bar{k}_V.$$

This relates the mean vertical flow to the mean vertical pressure gradient as in Darcy's equation $\overline{W} = -(\bar{k}_V/\mu)\partial \bar{p}/\partial z$ with an effective permeability

$$\bar{k}_V = \left\{ \frac{1}{h_R} \int_0^{h_R} \frac{dz}{k(z)} \right\}^{-1},$$

which, in a horizontally layered, binary medium with $k_{AQ} \gg k_{RL}$, reduces to

$$\bar{k}_V \approx k_{RL}/\beta. \tag{3.66}$$

where, again, β is the fraction of the total volume occupied by the retarding layers. Unlike (3.65), this result cannot be regarded as robust, since it depends crucially upon the assumptions of continuity and uniformity of the retarding layers and the absence of fractures traversing them. The flow is assumed to be uniformly vertical throughout the formation, an idealization may be satisfied only rarely in nature. It asserts that the average permeability is dominated by that in the regions of least permeability, that is, those that provide the greatest resistance to flow. In contrast, the minimum dissipation theorem implies that the flow will by-pass such regions wherever possible, and will follow a more circuitous but lower resistance path threaded among them. This is discussed more quantitatively below.

Be that as it may, if we take (3.65) and (3.66) at face value, they can be written in terms of the hydraulic conductivity (2.9). If the inverse of the hydraulic conductivity is called the "hydraulic resistivity," the expressions can be interpreted as asserting that in horizontal flow (along the layers), the effective conductivity is, simply, the average conductivity in the averaging volume, while in vertical flow (across the layers), the effective hydraulic *resistivity* is the average of the resistivity values over the averaging volume.

The expressions for the effective permeabilities averaged over the resolution scale are frequently used in numerical simulations of aquifer flow to define a permeability tensor

$$
\mathbf{k} = \begin{pmatrix} \overline{k}_{\mathrm{H}} & 0 & 0 \\ 0 & \overline{k}_{\mathrm{H}} & 0 \\ 0 & 0 & \overline{k}_{\mathrm{V}} \end{pmatrix},
\tag{3.67}
$$

so that when the small-scale layering is uniform, Laplace's equation for the pressure field is replaced by

$$
k_{\mathrm{H}}\frac{\partial^2 p}{\partial x^2} + k_{\mathrm{H}}\frac{\partial^2 p}{\partial y^2} + k_{\mathrm{V}}\frac{\partial^2 p}{\partial z^2} = 0.
\tag{3.68}
$$

As described earlier, in non-trivial solutions of Laplace's equation, the length scales l_{H} and l_{V} for flow patterns in vertical and horizontal directions are essentially the same, while in a uniformly layered medium represented by (3.68), the ratio of scales is

$$
l_{\mathrm{H}}/l_{\mathrm{V}} \sim (k_{\mathrm{H}}/k_{\mathrm{V}})^{1/2}.
\tag{3.69}
$$

This is frequently postulated to be quite large, possibly as much as one or two orders of magnitude.

The effective permeability expression (3.66), though commonly accepted and used in numerical simulations, is in fact a lower limit to the transverse permeability that depends crucially on the continuity and uniformity of the retarding layers throughout the entire flow domain. If a retarding layer is locally fractured, thinner, or coarser in texture, or disappears altogether, these "defects" may offer a more permeable pathway that is more circuitous, but still offers less flow resistance and therefore attracts flow from the direct path. It is unlikely that we will ever know the detailed topology of layering in an extensive flow domain with natural variability, but considerable insights (and robust results) on the effects of defects can often be gained from simple and approximate hydrological models based upon general principles such as the minimum dissipation theorem, that require less restrictive assumptions than those used above.

3.4.2 *Flow across layering with scattered fracture bands or gaps*

To examine the effect of scattered thinning or fracturing or other defects in retarding layers whatever the scale, consider the generic geometry illustrated schematically in Figure 3.16, in which aquifer layers of mean thickness h_{AQ} and permeability k_{AQ} are separated by retarding layers of average thickness h_{RL} and *much* smaller permeability k_{RL}. At various locations, these retarding layers are supposed much thinner than the mean, or locally absent, or fractured, etc., and these defects can allow local interstitial fluid to move from one aquifer layer to the next without crossing the intact and much less permeable parts of the retarding layer. The distances L separating the defects in adjacent retarding layers are expected to vary substantially in nature but in general, the separations may be much greater than the individual layer thicknesses. If large fracture zones traverse the whole structure, the permeability to vertical flow may be greatly increased, at least locally. The defects or gaps can be expected to be highly permeable, and offer less flow resistance than that for flow across the intact retarding layers where the permeability is low, and also for flow along the aquifers *between* separate defects despite its longer path length. We therefore consider situations in which $k_{AQ} \gg k_{RL}$ and the average horizontal distance between gaps or defects is large compared to the layer thicknesses, $\overline{L} \gg h_{AQ}, h_{RL}$, both possibly by orders of magnitude. Parameter ratios may vary among different sites by three or four orders of magnitude.

It is fairly obvious that in the geometry illustrated schematically in the top panel of Figure 3.16, with an even approximately horizontal mean pressure gradient, the average (resolution scale) permeability is close to that given by (3.64) and (3.65). Fluid moving along the aquifer layers can cross to an adjacent layer without much penalty in energy dissipation, as the center panel of Figure 3.16 suggests. On the other hand, in a discharge zone where the reduced pressure gradient is primarily *across* the striations as in the lowest panel, the connectivity provided by any gaps or defects would be expected to give an effective permeability significantly

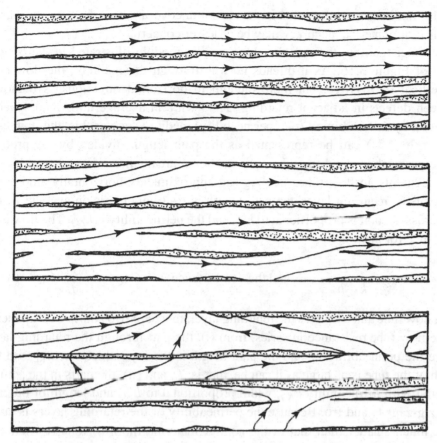

Figure 3.16. Flow patterns in a medium with less permeable lenses or finite-length layers with gaps. In generally horizontal flow, the fluid moves along the more permeable layers but may move among them without significant increase in flow resistance. If the flow is on average vertical, the effective mean permeability (3.76) involves the relative conductance of flow paths that traverse the less permeable layers and those that skirt around them.

greater than (3.66). A simple, approximate analysis using the concept of flow tube conductance (2.35), allows us to estimate the proportion of the total vertical flux that (i) flows through the gaps avoiding the retarding layers, and the fraction that (ii) seeps through them, despite their lower permeability. In this kind of physical model, we identify the parameters that define the flow and sort out their mutual interactions, but we ignore numerical coefficients that are expected to be of order unity. The solution will reveal how the flow properties change with different ratios in the individual defining parameters, but the end results will involve numerical proportionality constants that are of order unity but not defined more precisely. In many instances, the actual field values of the physical or geometrical parameters are also not known well, so that in the task of uncovering how the flow works, the sacrifice of numerical accuracy may be of lesser concern.

Consider the flow through the characteristic module illustrated in the bottom panel of Figure 3.16. As it threads its way through the medium, the flow may be vertical *on average*, but more random in horizontal directions as it moves between defects in adjacent retarding layers with an average horizontal spacing specified by $\overline{L}$. In a binary (two-permeability) medium such as this, the flow tube resistance (2.34) can be represented as the path length divided by the product of the permeability and the tube cross-sectional area, summed over the different flow segments. In flow tubes that zig-zag approximately horizontally among the segments of retarding layer, the path length in each module is on average $\overline{L}$, the tube cross-section is proportional to h_{AQ} and the permeability is k_{AQ}. The flow tube resistance for these is

$$R_{AQ} \sim \frac{\overline{L}}{k_{AQ} h_{AQ}} \quad \text{and the conductance} \quad C_{AQ} \sim \frac{k_{AQ} h_{AQ}}{\overline{L}}. \quad (3.70)$$

For a flow tube that traverses the retarding layer, there are two additive components of the flow tube resistance in series, from (i), the part through the retarding layer where the permeability is k_{RL}, the path length is the layer thickness h_{RL} and the width of the tube is as large as it can be, that is, $\overline{L}$, and (ii), the parts in the aquifer material with permeability k_{AQ}, width proportional to h_{AQ} and length of order $\overline{L}$ as it zig-zags to and fro. Because the permeability of the retarding layers is taken to be much smaller than that of the aquifer material, the flow tube resistance is dominated by the section through the retarding layer, so that

$$R_{RL} \sim \frac{h_{RL}}{k_{RL}\overline{L}} + \frac{\overline{L}}{k_{AQ} h_{AQ}} \approx \frac{h_{RL}}{k_{RL}\overline{L}} \quad \text{and} \quad C_{RL} \propto \frac{k_{RL}\overline{L}}{h_{AQ}}. \quad (3.71)$$

For flow paths "in parallel", i.e. that originate in one domain and travel by different paths to another, with given driving force, the flux in each path is proportional to the corresponding conductance, and as shown in Section 2.5, the total conductance is the sum of the conductances of the individual flow paths. Consequently, the fraction of the total flux that passes across the retarding layers and the fraction that avoids them are, respectively,

$$\frac{C_{RL}}{C_{AQ} + C_{RL}} \quad \text{and} \quad \frac{C_{AQ}}{C_{AQ} + C_{RL}}. \quad (3.72)$$

If the conductance of the flow tube C_{RL} that traverses the low-permeability, retarding layer is greater than that of the flow tube which skirts it, i.e. if $C_{RL} > C_{AQ}$, most of the flow does follow the shorter, lower-permeability path. From (3.70) and

(3.71), this is so when

$$L > \left(\frac{k_{AQ}}{k_{RL}}\right)^{1/2} (h_{AQ}h_{RL})^{1/2} \sim \left(\frac{k_{AQ}}{k_{RL}}\right)^{1/2} h, \qquad (3.73)$$

where h is the harmonic mean of the two sub-layer thicknesses. Note that the permeability ratio is assumed to be large, so that significant flow will traverse the less permeable layers when the average distance L between gaps is a sufficiently large multiple of the layer thickness, as one might have guessed. The overall permeability to flow across the laminations is then approximated by the classical (uniform layer) expression k_{RL}/β given below equation (3.66), where β is the fraction of the total section occupied by low-permeability layers. When the fractures or gaps are denser and the intervals between them are smaller, the inequality is reversed. Then

$$L \ll \left(\frac{k_{AQ}}{k_{RL}}\right)^{1/2} (h_{AQ}h_{RL})^{1/2} \sim \left(\frac{k_{AQ}}{k_{RL}}\right)^{1/2} h \qquad (3.74)$$

and the rising (or falling) fluid threads its way along the high-permeability pathways among the segments of the low-permeability layers. The overall permeability is proportional to the aquifer layer permeability k_{AQ}, with a geometrical proportionality factor that expresses the tortuosity of the pathways.

The pressure distributions associated with the zig-zag mean vertical flow pattern are of interest since their net effect determines the effective mean permeability for flow across the layers. In the bottom panel of Figure 3.16, the fraction $f = C_{AQ}/(C_{AQ} + C_{RL})$ of the total flux Q percolates laterally from one defect to the next one in the retarding layer above with transport velocity fQ/h_{AQ}, and, from Darcy's equation, this requires a pressure drop from one to the other such that

$$\frac{\Delta p}{L} = \frac{\mu}{k_{AQ}} \frac{fQ}{h_{AQ}}.$$

Thus

$$\Delta p = \mu QfL/h_{AQ}k_{AQ} = \mu Q \frac{1}{C_{RL} + C_{AQ}} \qquad (3.75)$$

after a little algebraic manipulation and use of (3.70). This is also the pressure difference across an individual retarding layer (within a numerical factor of about 1 or 2), so that the overall mean vertical pressure gradient is $\Delta p/(h_{RL} + h_{AQ})$. The relation between this and the overall mean transport velocity Q/L defines the effective mean permeability $\overline{k}$ for vertical overall flow in a layered medium with scattered gaps or fracture zones.

$$\frac{\mu}{\overline{k}} \frac{Q}{L} \approx \frac{\Delta p}{\Delta z} = \frac{\Delta p}{h_{AQ} + h_{RL}} \approx \frac{\mu Q}{h_{AQ} + h_{RL}} \{C_{AQ} + C_{RL}\}^{-1}.$$

Accordingly, the effective mean permeability for vertical mean flow is

$$\bar{k} \approx \left(\frac{h_{\mathrm{AQ}} + h_{\mathrm{RL}}}{L} \right) \{ C_{\mathrm{AQ}} + C_{\mathrm{RL}} \}. \tag{3.76}$$

The first factor in this equation is purely geometrical, the ratio of total thickness of a retarding layer/aquifer unit to the mean distance between defects. The second factor asserts that the mean permeability $\bar{k}$ is dominated by the medium with greater flow tube conductance or the smaller resistance (the aquifer material, as one might expect). With use of (3.27) and (3.28), the previous equation can be rewritten in an alternative form which offers additional physical insight:

$$\bar{k} \approx k_{\mathrm{RL}} \left(\frac{h_{\mathrm{AQ}} + h_{\mathrm{RL}}}{h_{\mathrm{RL}}} \right) \left\{ 1 + \frac{k_{\mathrm{AQ}}}{k_{\mathrm{RL}}} \left(\frac{h_{\mathrm{AQ}} h_{\mathrm{RL}}}{L^2} \right) \right\}. \tag{3.77}$$

The inverse of the second factor in this expression is simply the fraction of the total column occupied by the retarding layers, represented by β in (3.65) and (3.66). These first two factors in (3.77) can therefore be recognized as the approximation k_{RL}/β for uniform layering of indefinite length without defects. The third factor in (3.77) is the fractional increase in overall permeability produced by gaps or fractured regions, which involves the *large* permeability ratio between the aquifer and retarding layers and the generally *small* ratio of layer thicknesses to mean interval between defects.

Whether or not the defects are important in vertical overall flow depends on the competition between these two ratios and the transition involved in (3.73). For example, if the ratio of permeability in the aquifer and retarding layers is 10^4, and the mean interval between gaps is about 100 times the aquifer/retarding layer unit thickness, the mean permeability to vertical flow is only twice that in the defect-free case ($L \rightarrow \infty$), whereas if they occur three times more frequently, the mean permeability is about 12 times larger than given by the conventional expression (3.64).

3.4.3 Confining layers in a surface aquifer

On a larger scale, the dispersal of contaminants introduced into the groundwater across the water table of a surface aquifer may be influenced by the presence of one or more "confining layers," whose ability to restrict most of the infiltrating rainwater flow to relatively shallow depths is of interest. The extent to which this confinement occurs in a particular structure may not be obvious *a priori*. However, a simple analysis using the minimum dissipation principle indicates that the flow will be largely confined to the surface aquifer if the ratio of the confining layer thickness to aquifer length, though small, is at least an order of magnitude greater

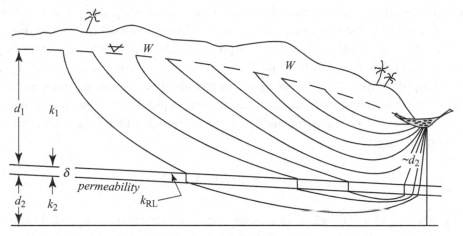

Figure 3.17. A surface aquifer with a single "confining" layer above the basement; the vertical scale is greatly exaggerated. From equation (3.78), effective confinement of surface contamination requires that the ratio of retarding layer permeability to aquifer permeability be much less than the ratio of retarding layer thickness to *aquifer length*, which usually is already small. Contamination introduced near the groundwater divide seeps into the lower aquifer and flushes much more slowly than that which enters at points closer to the discharge.

than the ratio of confining layer to aquifer permeability. Any contaminant that does leak through the confining layer originates near the groundwater divide as indicated schematically in Figure 3.17, although the leakage is distributed along most of its length. The upward flow to discharge is much more concentrated, as earlier examples have shown, with a width expected to be of the order of the aquifer depth, and the energy dissipation rate for the discharging flow upward across the retarding layer can be significant in the overall energy balance. A somewhat unexpected result from the analysis is the importance of the overall aquifer length L in determining the qualitative nature of the flow pattern. For given permeability parameters, depth of formation and layer thicknesses, it is shown below that when the aquifer length is relatively small, the flow may be almost entirely confined to the upper layer, although with the same vertical permeability structure in a *long* aquifer, the flow can circulate more freely through the retarding layer to greater depths.

The analysis proceeds as follows. We consider the family of kinematically possible flow paths from infiltration to discharge and estimate the dissipation rates in the various hydrologic regions of the overall flow pattern as a function of the fraction f of the total flux that crosses the retarding layer in order to find the one in which the total dissipation is least. In Figure 3.17, a two-dimensional aquifer is supplied by a uniform average infiltration rate W over its length from the groundwater divide at the left of the figure to a discharge strip along a stream bed, say, a distance L to the right. In this analysis, the flow is assumed two-dimensional. The aquifer has permeability k_1 to a depth d_1. Below this, there is a retarding layer of thickness δ and with a much smaller permeability, k_{RL}, followed by a second aquifer layer of thickness d_2 and permeability k_2 above an impermeable basement. Characteristically, the total depth of the aquifer is very small compared with its extent, L, and the permeability of the aquifers is much larger than that of the retarding layer:

$$d_1 + \delta + d_2 \ll L \quad \text{and} \quad k_1, k_2 \gg k_{RL}.$$

The general pattern of flow is obviously from the groundwater divide, with the transport velocity very nearly uniform with depth in each aquifer region and increasing linearly with distance from the groundwater divide. The upward discharge is concentrated within a distance of order d_2 of the far end of the aquifer system and the energy dissipation rate there turns out to be a significant part of the whole. From (2.38), the total dissipation per unit transverse distance of the aquifer is

$$\varepsilon_{TOT} = \mu \int (u^2/k)dA \approx \mu \sum (u^2/k)A_i, \tag{3.78}$$

where A_i represents the area and k the appropriate permeability of each of the four domains specified above: the two aquifer regions with the longitudinal velocity component generally much larger than the vertical component; the low-permeability confining layer in which the flow is directly across it, not along; and finally, the discharge region.

The dynamical characteristics of these regions are specified in Table 3.1. At any section a distance x from the discharge, the infiltration W provides a total horizontal flux Wx in the system, and let f represent the unknown fraction of the total infiltration (to be determined) that leaks into the lower aquifer. We call this the leakage fraction. The fraction $1 - f$ enters the system over the distance $(1 - f)L$ from the discharge, and remains in the upper aquifer throughout, moving laterally with the transport velocity $u_1 = Wx/d_1$.

The flow *across* the low permeability layer is distributed and downward in the infiltration region and equal to fW, but it is more concentrated and upward near the discharge, with transport velocity of order fWL/d_2. Because the dissipation is quadratic in the velocity and in spite of the smallness of the discharge area, the total dissipation here is larger by a factor $L/d_2 \gg 1$ than along all the rest of the confining layer. Consequently from (3.78), the total dissipation is, in essence, the sum of that in the flow along the upper aquifer, along the lower aquifer and

Table 3.1. *Flow characteristics in the various sub-regions of the aquifer*

	Upper aquifer	Retarding layer	Lower aquifer	Discharge
Longitudinal flux	$(1-f)Wx$	Negligible	fWx	(upward)
Longitudinal velocity	$(1-f)Wx/d_1$	Negligible	fWx/d_2	
Vertical velocity	$\sim W$	$-fW$	$< W$	$\sim fWL/d_2$
Flow resistance	$(R_{AQ})_1 = \frac{\mu L}{k_1 d_1}$	$R_{RL} = \frac{\mu\delta}{k_{RL}L}$	$(R_{AQ})_2 = \frac{\mu L}{k_2 d_2}$	$\frac{\mu\delta}{k_{RL}d_2}$
Area	Ld_1	$L\delta$	Ld_2	$d_2\delta$
Regional dissipation rate	$\frac{\mu}{k_1}\left(\frac{(1-f)W}{d_1}\right)^2 d_1 L^3$	$\sim \frac{\mu}{k_{RL}}(fW)^2 L\delta$	$\frac{\mu}{k_2}\left(\frac{fW}{d_2}\right)^2 d_2 L^3$	$\frac{\mu}{k_{RL}}\left(\frac{fWL}{d_2}\right)^2 d_2\delta$

Numerical coefficients of order unity and terms of relative order d/L in the accounting are omitted. See Figure 3.17 for the geometrical specifications. W is the mean infiltration rate, f is the fraction that crosses the retarding layer and enters the lower aquifer.

across the retarding layer in the discharge region, respectively,

$$\frac{\varepsilon_{TOT}}{\mu} \approx \left(\frac{(1-f)WL}{d_1}\right)^2 \frac{Ld_1}{k_1} + \left(\frac{fWL}{d_2}\right)^2 \frac{Ld_2}{k_2} + \left\{\frac{fWL}{d_2}\right\}^2 \frac{d_2\delta}{k_{RL}}.$$

According to the minimum dissipation theorem, among all kinematically possible flows satisfying given boundary conditions, the actual flow is the one in which the total dissipation is least. The fraction f of the total flux that infiltrates into the lower aquifer defines the family of conceivable flow patterns, and the member of the family that also satisfies the dynamical equations is that which provides the minimum total dissipation rate where $\partial\varepsilon_{TOT}/\partial f = 0$. We then differentiate the expression above, equate the result to zero and solve for f. After a little algebra, and with the recollection that

Flow resistance = (Viscosity) · (Path length)/(Permeability) · (Path width)

it is found that the leakage fraction f at the minimum value of ε_{TOT} is

$$f = \frac{(R_{AQ})_1}{(R_{AQ})_1 + (R_{AQ})_2 + R_{RL}}. \tag{3.79a}$$

This expression can be usefully interpreted in term of the flow tube conductances for the flow paths. For the pathways always above the confining layer, the conductance is $C_1 = 1/(R_{AQ})_1 \sim k_{AQ}d/\mu L$. The path traversing the confining layer contains, sequentially (i.e. in series), the two resistance components $(R_{AQ})_2$ along the aquifer and R_{RL} across it, with a total resistance $(R_{AQ})_2 + R_{RL}$ and conductance $C_2 = 1/[(R_{AQ})_2 + R_{RL}]$. Thus the leakage fraction can be written

$$f = \frac{1/C_1}{(1/C_2) + 1/C_1} = \frac{C_2}{C_1 + C_2}, \tag{3.79b}$$

which is the ratio of the conductance through the lower layer to the total conductance, a result in accordance with network theory and with intuition.

Note that the leakage fraction f is always non-zero; some of the infiltrating rainwater from a region close to the groundwater divide always seeps through to the deeper aquifer. *Deeper* streamlines originate from *closer* to the groundwater divide, as in Figure 3.17, and any contamination entering the water table near the here finds its way to the lower aquifer as it moves towards the discharge region. Because of the spatially uniform infiltration assumed in this example, the leakage fraction f also specifies the fractional distance from groundwater divide to discharge that provides the origin of the lower aquifer flow. Contamination flow paths that enter the system across the water table at points more remote from the groundwater divide than this and closer to the discharge, follow a shallower streamline that does not enter the lower aquifer at all, but discharges directly from the upper aquifer.

Effective confinement of surface-induced contaminant to the upper aquifer requires that the leakage fraction f, is small, or that $R_{RL} \gg (R_{AQ})_{1,2}$ and $C_1 \gg C_2$. In terms of the basic parameters

$$f \approx \frac{(R_{AQ})_{1,2}}{R_{RL}} = \frac{k_{RL}L}{k_{AQ}\delta} \ll 1$$

by a factor of at least 10. Alternatively, effective containment requires that

$$\frac{\delta}{L} \gg \frac{k_{RL}}{k_{AQ}}. \tag{3.80}$$

i.e. that the ratio of the retarding layer thickness to aquifer *length* must be at least an order of magnitude greater than the ratio of retarding layer permeability to aquifer permeability.

It is important to note that the leakage fraction depends not only on the permeability of the aquifers and the thickness of the retarding layer, *but also on the aquifer length*. In a short aquifer with a given vertical structure of layer thickness and permeability, the retarding layer might confine the flow almost entirely to the upper layer, while the same vertical structure in a much longer aquifer loses its ability to confine and allows the flow to be distributed throughout. The physical reason for this is that, as the aquifer length L increases from a relatively small value, the flow resistance of the upper aquifer *increases with the length* while that of the discharge flow across the retarding layer remains constant and therefore a smaller fraction of the whole. Ultimately, it becomes more energy-efficient for a greater part of the flow to divert across the "retarding" layer and take advantage of the low resistance flow path below.

For example, the surface aquifer in the Delmarva Peninsula, illustrated in Figure 3.6 (from Shedlock *et al.*, 1999, Bachman, 1984) consists of a major sand unit about 30 m thick, overlain by complex layering of some 5 m of much less permeable clay, silt and peat and another 10 m of sand. The permeability ratio k_{RL}/k_{AQ} appears generally to be about 10^{-4}. The overall aquifer consists of many small watersheds with groundwater flow paths that range from 300 m to about 5 km in length. For those with the shortest groundwater paths, $\delta/L \sim 10^{-2}$ so that the condition (3.80) is well satisfied and the leakage fraction is very small. The flow in these watersheds is indeed largely confined to the aquifers above the retarding or confining layers. For the longer aquifers having the same vertical structure but with longer flow path lengths $L \sim 5$ km, we have $\delta/L \sim 10^{-3}$ and (3.80) is only marginally satisfied, indicating rather poor confinement and about 10% leakage from the upper to the lower aquifer.

3.4.4 Mixing in more permeable lenses

The concentration of flow through lenses that are more permeable than the surrounding matrix enhances the lateral and vertical mixing of solutes. As fluid enters a lens, any pre-existing vertical gradient of an unsaturated solute is amplified in the convergence by a factor equal to the focusing ratio. Inside the lens, the longitudinal interstitial flow speed v increases by the same ratio; vertical concentration gradients are then very large and vertical dispersion inside the lens is greatly enhanced. If the flow distance l through the lens is sufficiently large, i.e. if,

$$l \geq \delta^2/\alpha_D$$

where δ is the thickness of the lens and α_D its dispersivity (2.80), the fluid in the lens may have become mixed before it emerges at the other end and spreads out vertically. The final vertical gradient in concentration of the solute in the fluid that has passed through the lens is then very much less than it had been initially.

For two-dimensional flow in a confined aquifer with total thickness d and permeability k_A as illustrated in Figure 3.18, the volume flux q is independent of distance along the aquifer, although the longitudinal pressure gradient and the partition of the flux among the lenses may vary. If lenses of permeability k_L occupy the amount δ of the aquifer thickness, then

$$q = \int u\,dz = u_L\delta + u_A(d - \delta)$$
$$= -(k_L\delta + k_A(d - \delta))v^{-1}\,(\partial p/\partial x) = \text{const.,}$$

(3.81)

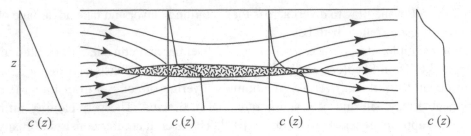

Figure 3.18. Isolated permeable lenses in a less-permeable matrix focus the flow and the convergence amplifies any vertical gradient in dissolved solutes, promoting accelerated vertical mixing.

where $k_L \gg k_A$. For increasing total lens thickness δ, more of the total volume flux passes through the lenses, the longitudinal pressure gradient decreases and the flow *outside* the lenses is correspondingly slower, as is evident from the streamline spacing in Figure 3.18.

In a surface aquifer, the longitudinal pressure gradient is maintained by the water table slope and, from (3.23),

$$q = -C\nabla_H\zeta = -C\frac{\partial\zeta}{\partial x}$$

in two-dimensional flow, where C is the vertically integrated hydraulic conductivity (3.24). Physically, the flux q is determined by the upstream infiltration, and the water table slope adjusts to variations in C along the flow path, being smaller when the lenses occupy more of the section. If the interstitial fluid near the water table contains chemically passive contaminants, the convergence and intermixing inside the lens allow the contaminant to be distributed vertically down-stream of the lenses.

3.5 Fracture–matrix or "crack and block" media

Fractures are almost ubiquitous in natural rock formations. The marble mountains of Carrera in Italy have been famous since antiquity for the huge blocks of pristine white marble that they yield, their rarity making them noteworthy. More commonly, rock formations and lithified layers are fractured on scales of order meters or less, as a legacy of stresses from tectonic movements or thermal gradients over time past. The hydrological properties of fracture–matrix media have many interesting characteristics that are becoming increasingly important in questions of contaminant transport and mineral deposition.

As mentioned earlier, a fracture plane should not be considered as a gap between parallel plane rock surfaces. It is, more realistically, an approximately plane network of intersecting ribbon-shaped pathways, separating discrete close-contact

areas of adjacent matrix blocks. As with a simple "sandbank" medium, averages
are defined over scales large compared with the microscale, which here is the representative block size, or distance between fractures. This scale is assumed to be
small compared with the depth scale of the formation but this may not always be so;
the alternative is to consider the flow in separate and individual fractures interacting
with the matrix. If λ represents the mean total length of the ribbon pathway network
intersected per unit area of a slice through the medium with dimension [length]$^{-1}$,
then the fracture porosity is $\phi_F = \lambda \bar{\delta}_F$, as in (2.3), where the mean gap width is
$\bar{\delta}_F$. Characteristically, $\delta_F \sim 10^{-5} - 10^{-4}$ m (10–100 microns), with occasionally
much larger values, so that with a moderate density of active fracture segments,
$\lambda \sim 1$ m/m^2, say, the fracture porosity is also $\phi_F \sim 10^{-5} - 10^{-4}$, very much smaller
than the matrix porosities in Chapter 2.

In contrast, the fracture permeability is often much *greater* than the bulk permeability of many un-fractured rocks. If the fracture geometry is idealized as a set of
parallel ribbon-shaped apertures of thickness δ_F, the local Poiseuille flow solution
(see Batchelor, 1967) gives for the fracture permeability,

$$k_F = \tfrac{1}{12}\phi_F \delta_F^2 = \tfrac{1}{12}\delta_F^3 \lambda \qquad (3.82)$$

when the channels are aligned in the direction of the pressure gradient. If their
directions differ by an angle θ, an additional factor $\cos^2 \theta$ is involved in (3.82),
because (a), the component of the pressure gradient is smaller by $\cos \theta$ and (b), the
path length per unit slice normal to the pressure gradient (and flow resistance) is
greater by $(\cos \theta)^{-1}$. This implies that in a relatively thin layer containing multiple
transverse fractures with random orientations in the plane of the layer, as shown
in Figure 3.19, the permeability is anisotropic, but not strongly so. For transverse
flow, the fracture paths are all in the flow direction as in (3.82), while for flow along
the layer the angle θ is distributed randomly and the $\cos^2 \theta$ dependence averaged
over all directions reduces the permeability for this flow direction, but only by
one-half. More realistically, if the aperture widths vary and the directions of active
fracture flow paths are distributed randomly in *all* directions, it can be shown that

$$k_F = \phi_F \overline{\delta_F^2}/36 = \overline{\delta_F \delta_F^2}\lambda/36, \qquad (3.83)$$

provided the connectivity of the fracture flow paths is sufficient to avoid significant
choking by local constrictions. Little data are available on distributions of fracture
aperture widths, but if the larger apertures have $\bar{\delta}_F \sim 10^{-4}$ m, and the length
of active aperture per unit area $\lambda \sim 1$ m^{-1}, the fracture permeability is about
3×10^{-14} m^2, the same order of magnitude as found in field measurements on
granite and metamorphic rocks, given in Table 2.1. This provides some support to
Brace's (1980) conjecture that at least some of the low permeability values inferred

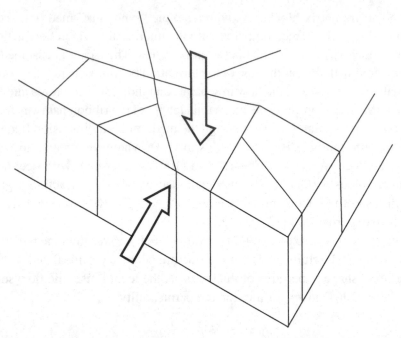

Figure 3.19. The overall permeability of a thin fractured layer is generally anisotropic, with the permeability for flow across the layer being about twice that for flow along it.

from field measurements in tight rocks are expressions of k_F, rather than matrix permeability k_M.

In nature, the *variations* in fracture path aperture δ_F are certainly considerable and because of the cubic dependence in (3.83) the distribution of flow is expected to be something qualitatively like log-normal, concentrated disproportionately in the relatively few widest pathways with the much more numerous, narrower pathways contributing much less. Measurements on the distribution of flow paths by Abelin *et al.* (1991) and Birgersson *et al.* (1995) in a fracture zone in granite and Salve and co-workers (2001, 2003) in unsaturated, non-welded tuff showed this channeling very clearly, with most of the flow taking place preferentially along relatively few pathways. There is no reason to believe that this property is specific to those particular sites. Casual observations of the distribution of stalactites in limestone caverns suggest also that the thicker stalactites, associated with the largest fluid pathways, are often separated by many smaller and incomplete ones along the lines of roof fractures.

In hydrodynamically driven flow, such as that in a surface aquifer, the overall mean pressure gradient is supplied by variations in water table elevation. The pressure gradient inside any particular block must be influenced by its configuration, the connectivity of the surrounding fractures, etc., and the flow pattern is unlikely to be straight through the block, from one side to the other. Nevertheless, the minimum dissipation theorem assures us that the flow within the block will be as efficient as possible, converging towards the interfaces that have the largest connected apertures in the surrounding fracture, and avoiding the more resistive regions. Consequently, the geometry of the contact areas is less pertinent to the overall flow (on the scale of our averaging volume) than parameters such as λ, the active connected area of ribbon pathways per unit volume of the fabric (c.f. Section 2.2) and the fracture gap parameters above.

3.5.1 *Reservoirs and conduits*

Several representative ratios are illuminating in understanding the co-existing and interacting flows in the saturated fracture network and matrix blocks. Assume here that the matrix blocks have porosity $\phi_M = 10^{-1}$ and permeability $k_M = 10^{-14}$ –10^{-16} m^2, while the fracture network has porosity $\phi_F \sim 10^{-4}$ and permeability $k_F = 10^{-13}$–10^{-14} m^2. With these figures, we have the following.

(1) The ratio of fluid storage volume in the matrix blocks to that in the fracture network, the ratio of the porosities, is $\phi_M/\phi_F \sim 10^3$. There is about one thousand times more fluid in the matrix blocks than there is in the fracture network. The size of this ratio is at the heart of the behavior of fracture–matrix flow, also called double porosity flow.

(2) The ratio of transport velocities or volume fluxes through the fracture network to those in the matrix blocks for a given pressure gradient or buoyancy force is generally quite large: $u_F/u_M = k_F/k_M \sim 10-10^3$.

(3) Because of the smallness of the porosity ratio, the ratio of interstitial fluid velocity in the fractures, $\bar{v}_F = u_F/\phi_F$, to that in the matrix blocks, $\bar{v}_M = u_M/\phi_M$ is *very* large, by an additional two or three orders of magnitude:

$$\frac{\bar{v}_F}{\bar{v}_M} = \left(\frac{k_F}{k_M}\right)\left(\frac{\phi_M}{\phi_F}\right) \sim 10^3 - 10^6. \tag{3.84}$$

Fluid idles in the matrix blocks but moves a thousand to a million times faster in the fracture network.

(4) If l_B represents the characteristic block size, the time taken for fluid elements to move through an individual block (the exchange time) is $l_B/\bar{v}_M$, and reciprocal of this is the exchange rate $E = \bar{v}_M/l_B$, an important parameter for these flows.

(5) The characteristic distance λ_F that a fluid element moves along the fracture network before being absorbed into a matrix block is large compared with the block size: $\lambda_F = (k_F/k_M)l_B \sim (10-10^3)l_B$. This is not immediately obvious, but is established in (3.96) below.

These ratios are central to understanding the distributions of dissolved solutes, both chemically active and passive, in fracture–matrix media. They emphasize that in a typical saturated fracture–matrix medium, the matrix blocks provide the dominant storehouses for water and dissolved solutes, while the fractures provide the dominant conduits for flow. The flow through the matrix blocks occurs throughout most of the space, but is sluggish, and generally provides less volume transport than the much higher-speed fracture flow through a tiny fraction of the domain. Even in the fractures, though, the flow Reynolds number is small, so that the basic Darcy balance persists between pressure gradient and viscous resistance, and equations can be written separately for the co-existing fracture networks and the matrix blocks:

$$u_M = -\frac{k_M}{\mu}\nabla p, \qquad u_F = -\frac{k_F}{\mu}\nabla p. \tag{3.85}$$

Although water is exchanged between the fractures and the matrix blocks, the incompressibility condition assures that fluid leaving a fracture or matrix block is simultaneously replaced by an identical volume entering. *If the fluid entering and that leaving the blocks are chemically indistinguishable*, there is little point in the separation (3.85), and the combined transport velocity is

$$u = u_M + u_F = -\frac{(k_M + k_F)}{\mu}\nabla p, \tag{3.86}$$

with the effective permeability for the total flow being the sum of the individual permeabilities of the blocks and the fracture network in parallel. In considerations of solute transport and reactions, however, the matrix and fracture flows must be considered separately as shown below, with provision for their mutual interactions.

All this has been concerned with fluid-saturated media. Above the water table, the matrix is unsaturated with both air and water in the void spaces. The technological questions involved in the long-term storage of radioactive wastes in underground repositories located in an arid climate, such as Nevada, have stimulated many important contributions to the understanding of water and solute transport in unsaturated fracture–matrix media. Some interesting phenomena might be noted briefly. In an unsaturated hydrophilic fracture–matrix medium such as those which surround the Nevada repository sites, capillarity draws water into the finer-scale pores, leaving the fracture network to provide pathways for moist air flow and obstacles to water flow from one matrix block to the next. The blocks can lose water by evaporation, with subsequent diffusion to neighboring fractures and advective transport through the open fracture pathways. Occasional infiltration from rainfall produces gravity-driven downward flux through the fracture network, generally in the form of fingers, that diminish with depth as a result of "imbibition" or absorption into the adjacent matrix blocks. Salve and co-workers (2001, 2003) and others have made interesting and important field measurements of these processes. The basic physics is not entirely understood, and little quantitative testing of theoretical results has yet been possible, so that most exploratory numerical models contain substantial empiricism. Extensive numerical simulations by Preuss *et al.* (1997) illustrate the variety of physical processes already identified and their interactions that need to be considered.

3.5.2 Transport of passive solute in co-existing fracture and matrix block flows

If a medium is saturated with fluid containing contaminants or reactants in solution, these can be exchanged between the fracture network and the matrix blocks, where they are advected at very different rates. Models to describe the co-existing, two-phase immiscible flows were used many years ago by Muscat and Meres (1936) and by Leverett (1939), while Barenblatt and his colleagues have developed 'double porosity' models in a variety of important applications to hydrology (1960) and the physics of oil recovery (1963, 1990). As usual, we consider averages over volume elements large compared with the characteristic block size or distance between fractures. The primary concept involves *separate* flow averaging of the fluid velocities, solute concentrations, etc., over the fracture network and over the matrix pores, allowing for interchange between them. This notion of co-existing and interacting fracture and matrix block flows has many interesting applications.

First, let us consider the movement and dispersion of passive solutes in an aquifer with this "crack-and-block" structure. The transport velocities for the matrix and fracture networks are as specified in equation (3.85). The conservation equation for a chemically passive solute, as specified in general by equation (2.50), is to be applied separately to the fracture network and the fabric of matrix blocks.

Inside any individual matrix block, the solute balance per unit volume of fabric is

$$\phi_M \frac{\partial c_M}{\partial t} + \mathbf{u}_M \cdot \nabla c_M = \phi_M D \nabla \cdot (\nabla c_M), \tag{3.87}$$

where D is the internal solute diffusivity. The volume integral of this equation over the block can be written as

$$\phi_M \int (\partial c_M / \partial t) dV + \int c_M \mathbf{u} \cdot d\mathbf{S} = \phi_M D \int (\partial c_M / \partial n) \cdot d\mathbf{S},$$

with use of the incompressibility condition (2.8) and the divergence theorem. This expresses the rate of change of total solute in the block in terms of the total flux of solute and the diffusion of solute across the block surface. The first term can be written as $\phi_M (\partial c_M / \partial t) V$ where $V \sim l_B^3$ is the volume and c_M (here) is the mean interstitial concentration of solute in this particular matrix block. The second integral, over the surface of the block, expresses the net advection of solute out of the block, with the flow entering from the adjacent fracture with the concentration c_F, say, and leaving with concentration c_M to rejoin the fracture flow. This can similarly be approximated as $\phi_M \bar{v}_M (c_M - c_F) l_B^2$, in which the cross-sectional area is represented as l_B^2. The last term expresses the net flux of solute in and out of the block by dispersion across the surface, but since the scale of the internal pores is much smaller than the block size, this effect can be neglected compared with advection. If we express the previous equation in these terms and average over all the blocks in our resolution volume, the solute balance in the matrix blocks reduces to

$$\phi_M \frac{\partial c_M}{\partial t} = -\phi_M E (c_M - c_F), \qquad (3.88)$$

where $E = \bar{v}_M / l_B$ is the exchange rate, the inverse of the mean time for fluid elements to move through blocks of typical size l_B. The exchange rate is not an intrinsic property of the flow, but a useful parameter of solute advection through matrix blocks.

In the solute balance for the fracture network, dispersion of solute in the convoluted pathways among the matrix blocks is much more important than it is within the blocks themselves. The dispersion term in (2.50) should generally be retained in the form $\phi_F D \nabla^2 c_F$, where the diffusivity $D = \bar{v}_F \alpha_D \sim \bar{v}_F l_0$. In this expression, α_D is the dispersivity of the fracture network (c.f. Section 2.10), which has the physical dimensions of (length) and which is expected to be a moderate multiple of (perhaps 2 to 5 times) the block size. The random fluid velocities involved in the dispersion are expected to be proportional to $\bar{v}_F$, the interstitial fracture flow *speed*, whatever its direction. The rate of increase of solute in the fractures per unit volume of the fabric is equal to the rate of decrease in the matrix blocks, as given by the term on the right of the previous equation, so that the solute balance in the fracture network is

$$\phi_F \frac{\partial c_F}{\partial t} + \phi_F \bar{\mathbf{v}}_F \cdot \nabla c_F = \phi_M E (c_M - c_F) + \phi_F \bar{v}_F \alpha_D \nabla^2 c_F. \qquad (3.89)$$

Note that the matrix porosity ϕ_M is involved also in the exchange term of the fracture network balance. Addition of the last two equations gives the overall solute balance,

$$\frac{\partial}{\partial t} (\phi_F c_F + \phi_M c_M) = -\phi_F \nabla \cdot (\bar{\mathbf{v}}_F c_F) + \phi_F \bar{v}_F \alpha_D \nabla^2 c_F, \qquad (3.90)$$

in which the rate of change of total solute per unit volume of the fabric is given as the divergence of the solute flux and dispersion in the fracture network alone. The matrix blocks provide for the storage of solute, the fracture network provides for its transport and allows for its dispersal.

The coupled equations (3.88) and (3.89) can be expressed a little more conveniently as

$$\frac{\partial c_M}{\partial t} = -E(c_M - c_F), \tag{3.91}$$

$$\frac{\partial c_F}{\partial t} + \bar{v}_F \cdot \nabla c_F = \frac{\phi_M}{\phi_F} E(c_M - c_F) + \bar{v}_F \alpha_D \nabla^2 c_F, \tag{3.92}$$

where $\phi_M/\phi_F \gg 1$, α_D represents the fracture network dispersivity and the incompressibility condition for the fracture fluid, $\nabla \cdot v_F = 0$, has been used. The physical balances involved in fracture–matrix solute transport are dominated by the largeness, in one guise or another, of the ratio $\phi_M/\phi_F \sim 10^3$.

3.5.3 A passive contaminant front in a fracture–matrix aquifer

Suppose that, as a result of some local event, solute-laden water begins to infiltrate laterally from a source region into an initially pristine aquifer, in which uniform internal flow extends along the length of the domain from source to discharge. The leading edge of the contamination forms a front that moves down-stream at a mean speed that is very much smaller than the interstitial fluid velocity in the fractures, because at any instant, most of the fluid is contained in the blocks and moving only slowly. We consider here the advection and exchange processes of a solute that is chemically passive, deferring to the next chapter the spatial characteristics of patterns of dissolution, deposition, sorbtion and chemical reaction that leave their imprint in the fabric.

The initial stage of contaminant dispersal is a relatively brief period less than the exchange time $t < l_B/\bar{v}_M = E^{-1}$. Suppose that initially pristine water enters the aquifer across the vertical plane $x = 0$, say, but from time $t = 0$, it contains a water-soluble but chemically inert contaminant with concentration $c_F = c_0$, say. This travels rapidly along the fracture network, gradually seeping into the matrix blocks and at the same time being diluted by cleaner water leaving the blocks downstream, to join the fracture flow. During the time interval of this initial pulse, the solute concentration in the blocks is still small $c_M \ll c_F$, and if for the moment we neglect dispersion of the solute about the mean fracture velocity by the convolutions of the fracture pathways, equation (3.92) reduces to

$$\frac{\partial c_F}{\partial t} + \bar{v}_F \frac{\partial c_F}{\partial x} \approx -\frac{\phi_M}{\phi_F} E c_F = -\frac{\bar{v}_F}{\lambda_F} c_F, \text{ say,} \tag{3.93}$$

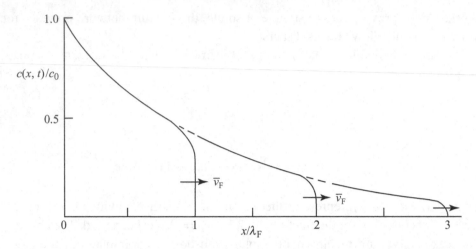

Figure 3.20. A pulse of passive contaminant in a uniform flow across the boundary of a fracture-matrix medium at $x = 0$ produces a front that moves with the interstitial (fracture) fluid and attenuates over the length scale λ_F (3.94) as solute leaks from the fractures into the adjacent blocks.

in which $\bar{v}_F$ is constant and the length

$$\lambda_F = \frac{\phi_F \bar{v}_F}{\phi_M E} \tag{3.94}$$

is a small fraction ($\phi_F/\phi_M \sim 10^{-3}$) of the distance solute moves along the fractures in the exchange time E^{-1}. The first two terms in (3.93) represent (c.f. (2.11)) the time rate of change of c_F following the mean fracture-fluid motion, and the last term expresses how the concentration changes as it moves along by fluid exchange with the fabric blocks, with fluid of concentration c_F entering from the cracks and still-uncontaminated fluid leaving. The differential equation has the solution

$$c_F(x, t) = c_0 F(x - \bar{v}_F t) \exp(-x/\lambda_F), \tag{3.95}$$

as can be established by substitution. The function $F(..)$ in this solution is arbitrary as far as the equation is concerned, and is determined by the time history of the infiltrating solute concentration, $c_F(t)$, at the source $x = 0$. If the solute (or contaminant) with concentration c_0 appears suddenly at time $t = 0$ and then remains constant, the distribution $F(x - \bar{v}_F t) = 0$ for $x > \bar{v}_F t$ and $= 1$ when $x < \bar{v}_F t$. The factor $F(..)$ thus represents a step moving outward along the aquifer as a wave with speed $\bar{v}_F$, while the final factor shows that the step size attenuates with distance on the scale λ_F, as indicated in Figure 3.20.

The exponential attenuation length in this initial stage solution can be interpreted in several useful ways. First, from the solution, λ_F is the characteristic distance that a fluid element, identified in this case by contaminant, moves along the fracture network before being absorbed into a matrix block. Secondly, from the definition of the exchange rate $E = \bar{v}_M/l_B$ and with use of (3.94) and the ratios given earlier in this section, we can infer its characteristic magnitude:

$$\lambda_F = \frac{\phi_F \, \bar{v}_F}{\phi_M \, \bar{v}_M} l_B = \frac{u_F}{u_M} l_B = \frac{k_F}{k_M} l_B \sim (10 - 10^3) l_B, \tag{3.96}$$

i.e. a moderate-to-large multiple of the block size. In a fracture–matrix medium, a contaminant "front" may be very diffuse!

The duration of the initial infiltration pulse is very short. When $t > \lambda_F/\bar{v}_F$, the step has moved out of the region of interest, the factor $F(. .) = 1$ in (3.95) and the subsequent changes in the distribution of contaminant in the matrix blocks are much slower. Continued solute inflow at the interface with the distributed seepage gradually increases the solute concentration there. Fluid entering the blocks does so at the incoming concentration $c_0 \exp(-x/\lambda_F)$ and leaves with concentration c_M, so that from (3.91)

$$\frac{\partial c_M}{\partial t} + E c_M = c_0 \exp(-x/\lambda_F)$$

and

$$c_M = c_0 \exp(-x/\lambda_F)\{1 - \exp(-Et)\}. \tag{3.97}$$

After a time $E^{-1} = l_B/\bar{v}_M$, the exchange time for the matrix blocks, those near the entry plane are becoming saturated with the solute-bearing incoming water, so that $c_M \sim c_F \sim c_0 \exp(-x/\lambda_F)$. With continued solute inflow, the saturated region gradually extends further along the flow streamlines, forming a "front" of thickness λ_F given by (3.96), separating the contaminated region from the pristine region ahead. Near the leading edge of the "front," blocks may be contaminated only along the fracture walls, while near the trailing edge contamination may have advected and dispersed more ubiquitously throughout the blocks. Both ahead of, and behind the front, the solute loads in the fractures and the blocks are in equilibrium, zero ahead and c_0 behind. Only inside the front, over a distance of order λ_F, is $c_F > c_M$, transferring solute from the fractures to the matrix.

The speed of advance V of this front through the medium can be established simply from the overall solute balance (3.90). With a uniform stream in the x-direction, the distributions c_F and c_M are each functions of $(x - Vt)$, so that $\partial/\partial t = -V\partial/\partial x$, and (3.90) reduces to

$$-V\frac{\partial}{\partial x}(\phi_F c_F + \phi_M c_M) + \bar{v}_F\frac{\partial}{\partial x}(\phi_F c_F) = \phi_F \bar{v}_F \alpha_D \frac{\partial^2 c_F}{\partial x^2}, \tag{3.98}$$

On integration of this across the front, the diffusion term on the right vanishes since the concentration gradients are zero both ahead of, and after the front, and because

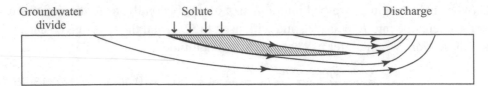

Figure 3.21. Cartoon of the distribution of passive solute (hatched) following injection maintained over a distributed surface source. The active contaminant front is parallel to the water table (approximately horizontal), moves downward and towards discharge; dispersion is neglected.

$c_M = c_F = 0$ ahead and $c_M = c_F = c_0$ behind it. The result is that

$$V = \frac{\phi_F \bar{v}_F}{\phi_M + \phi_F} \approx \frac{\phi_F \bar{v}_F}{\phi_M}, \qquad (3.99)$$

since $\phi_M \gg \phi_F$. This result can also be obtained without calculation. In a frame of reference moving to the right with the speed V of the front, the flow and solute concentration fields are steady. The flux of solute in the blocks and fractures being carried to the left by the motion of the medium in this frame is $(\phi_M + \phi_F)V c_0$, while that being carried to the right in the fractures by the fluid motion *relative to the medium* is $\phi_F \bar{v}_F c_0$. Since there is no accumulation of solute in the frontal region, these two must balance, so that (3.99) follows. Note that, unlike fronts of passive contaminants in a simple "sandbank" medium, which move at the interstitial fluid velocity, those in a fracture–matrix medium move much more slowly, because at every instant, most of the fluid is relatively stagnant in the matrix blocks.

3.5.4 Distributed solute entering across the water table

Suppose that infiltration across the water table into an initially pristine fracture–matrix aquifer becomes contaminated uniformly over a substantial area, possibly by regional application of excessive fertilizer that continues at a constant rate. The mean flow pattern with averaging volumes large compared to the characteristic distance between fractures is illustrated schematically in Figure 3.21, though most of the flow actually occurs in relatively few fracture conduits. In the mean, the solute follows the streamlines except for dispersion, and moves uniformly downward at a linearly decreasing rate, so that if the aquifer permeability and the average infiltration rate are both uniform, the advancing contamination front remains approximately parallel to the water table as it moves downward and towards the discharge region. The time derivative following the motion in equation (3.91) is now the derivative along the streamlines, which intersect the concentration front at a generally small angle. Although the mean solute concentration gradient near the front is vertical, the flux of contaminant through the fracture network, $(\bar{v}_F)c_F$, dips only slightly from the horizontal, as described in Section 3.1 and the aquifer measurements of Section 3.3. Along any given mean streamline through the front, the distance over which solutes in the fracture flow leak into more pristine matrix blocks is much greater than the thickness of the front as measured in the vertical direction.

The advection term in the Eulerian equation (3.92) then involves only the relatively small vertical component of the mean interstitial velocity in the fracture network, $(\bar{v}_F)_z = -(W/\phi_F)(1 + z/D) = -(\bar{v}_F)_0(1 + z/D)$, say, where $(\bar{v}_F)_0$ (>0) is the mean interstitial fluid fracture velocity associated with the infiltration at the water table. If $W \sim 1$ m/yr, and the porosity of the fracture conduits network $\phi_F \sim 10^{-3}$, then $(\bar{v}_F)_0 \sim (2 - 3\text{m})$/day, decreasing linearly with depth. In contrast, the *dispersion* of solute in (3.92) produced by the interstitial flow around the blocks, is proportional to the total mean interstitial fracture fluid speed, which is approximately uniform in depth, predominantly along the aquifer and approximately equal to Wl/D, where l is the distance from the groundwater divide. Consequently, the effective solute diffusivity for this flow can be taken as

$$\alpha_D(W/\phi_F)(l/D) = \alpha_D(\bar{v}_F)_0(l/D),$$

where α_D is the vertical dispersivity in essentially horizontal flow. (For dynamical reasons, this is very small or zero in the sandy aquifers discussed in Section 3.3.)

In place of (3.92), the downward propagation of the contaminant front is described by

$$\frac{\partial c_F}{\partial t} - (\bar{v}_F)_0\left(1 + \frac{z}{D}\right)\frac{\partial c_F}{\partial z} = \frac{\phi_M}{\phi_F}E(c_M - c_F) + (\bar{v}_F)_0\alpha_D\frac{\partial^2 c_F}{\partial z^2}, \quad (3.100)$$

where $0 \geq z \geq -D$ and the negative sign before the second term reflects the downward direction of the advection. This equation differs from (3.93) in several ways. Most significantly, the advection velocity, the coefficient of the space derivative term is no longer constant, but varies linearly with depth between $(\bar{v}_F)_0$ at the upper surface $z = 0$ (the water table) to zero at the base, $z = -D$. After a time of order E^{-1} the contamination front propagates downward at the speed

$$V = \frac{W}{\phi_F}\left(1 + \frac{z}{D}\right), \quad (3.101)$$

which can be compared with (3.97). The speed at which the front propagates downward through the fractures is initially large but decreases with depth until, when it is close to the base, it is advancing so slowly that the vertical dispersion of solute in the longitudinal motion close to the bed may begin to dominate the distribution.

Patterns of flow

3.6 Flow transients

3.6.1 Diffusion of pressure

Following a local, impulsive disturbance in a permeable rock medium, both the stress field in the surrounding rock and the pressure in the interstitial fluid are altered. The compressibility of most rock material is only about 3–4% of that of water, a useful tabulation having been given by Daly (1951). An approximate estimate of the ratio can be made simply by comparing the speeds of propagation c_P and c_S of seismic P- and S-waves in rocks, to the speed of sound in water. In rocks, the elastic compressibility modulus is

$$e_c = \left\{ \rho \left(c_P^2 - \tfrac{4}{3} c_S^2 \right) \right\}^{-1}$$

where, from seismic measurements, $c_P \sim 6$ km/s and $c_S \sim 3$ to 4 km/s in sandstone or limestone. The compressibility e_W of water is $(\rho_W c^2)^{-1}$, where the speed c of sound in water is about 1.5 km/s. These figures give a ratio e_W/e_c of about 30 – water is much more compressible than rocks are. Care must be taken, particularly in the case of saturated clay minerals, to distinguish between the compressibility of the matrix as a whole and that of the clay mineral itself. When the porosity is relatively large and the rigidity of the solid is small, the compressibility of the overall matrix is dominated by that of the interstitial water. With this proviso, then, to a reasonable accuracy the rock material can be regarded as incompressible and its porosity ϕ as constant, whereas the interstitial fluid seeks to expand or contract following pressure changes.

In the mass conservation equation (2.6) the changes in fluid density with time are now important, but we can neglect the small nonlinear product of density and velocity fluctuations, and without the source term it becomes

$$\phi \frac{\partial \rho}{\partial t} = -\rho_0 \nabla \cdot \mathbf{u}, \tag{3.102}$$

where ρ_0 is the mean water density. Since the thermal capacity of the rock matrix is usually much greater than that of the interstitial water, the adjustment process in the interstitial fluid is close to isothermal. Density and pressure changes are related by the isothermal sound speed in water, c_I, which in liquids is very close to the usual adiabatic sound speed:

$$\partial p = c_I^2 \partial \rho \quad \text{and} \quad \partial \rho = c_I^{-2} \partial p.$$

Consequently, from (3.102),

$$\phi \frac{\partial p}{\partial t} = -\rho_0 c_I^2 \nabla \cdot \mathbf{u}. \tag{3.103}$$

Suppose, for simplicity, that the permeable medium is homogeneous and locally isotropic and the fluid obeys Darcy's equation $\mathbf{u} = -(k/\mu)\nabla p$ so that the fluid divergence

$$\nabla \cdot \mathbf{u} = -(k/\mu)\nabla^2 p.$$

Equation (3.103) then becomes

$$\frac{\partial p}{\partial t} = \frac{kc_{\mathrm{I}}^2}{\nu\phi}\nabla^2 p = \kappa_{\mathrm{P}}\nabla^2 p, \text{ say,} \tag{3.104}$$

where ν is the kinematic viscosity of the interstitial water. This is a classical unsteady heat conduction or diffusion equation with the same structure as (2.48) or (2.50) without the advection and source terms. The coefficient in front of the Laplacian, the "pressure diffusivity" is

$$\kappa_{\mathrm{P}} = \frac{kc_{\mathrm{I}}^2}{\nu\phi}. \tag{3.105}$$

This is analogous to the thermal diffusivity in Fourier's law and in the heat conduction equation, which expresses the capability of the medium to transmit heat and to smooth out temperature variations. Similarly, the pressure diffusivity expresses the ability of the pressure in the interstitial fluid to re-establish its equilibrium following a change in the distribution of fracture and pore spaces. The structure of the expression (3.105) indicates the properties of the medium involved in the readjustment. In the numerator, a greater permeability k (with a given porosity) allows the interstitial fluid to move more readily through the pores, and the volumetric elasticity allows it to expand or contract in response to pressure variations. Both factors increase the pressure diffusivity. In the denominator, a larger viscosity reduces the freedom to move through the pores and a larger porosity (with a given permeability) means that more fluid must be moved in the readjustment, both reducing the pressure diffusivity. For water with $c_{\mathrm{I}} \sim 1.4 \times 10^3$ m/s and $\nu \sim 10^{-6}$ m^2/s with $\phi = 0.2$ and $k = 10^{-13}$ m^2, it is found that $\kappa_{\mathrm{P}} \sim 1$ m^2/s. By comparison, the thermal diffusivity of saturated rocks is very much smaller, about 10^{-7} m^2/s; pressure diffuses much more rapidly than heat.

The analogy with heat conduction is very useful. The way that the internal pressure field changes in the interstitial fluid in response to an impulsive change in boundary pressure (resulting from a seismic event, perhaps) is exactly analogous to the way that the internal temperature field would change in response to an impulsive temperature change at the boundary, with the thermal conductivity equal to κ_{P}. In a uniform "sandbank" medium, a pressure disturbance diffuses through a characteristic distance d in time t given by

$$t = d^2/2\kappa_{\mathrm{P}}, \tag{3.106}$$

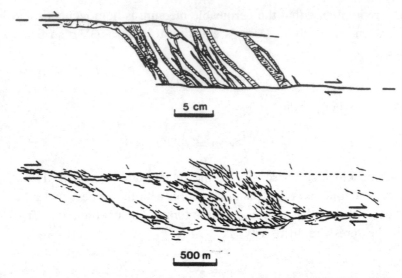

Figure 3.22. Examples of dilatational fault jogs, from Sibson (1987). At the top, a small-scale jog in sandstone has been infilled with locally fibrous quartz. Below is a segment of surface rupture traces of the 1968 Dasht-e-Bayaz earthquake in Iran, after Tchalenko and Ambraseys (1970). Note the difference in scales.

as does heat, with the appropriate thermal diffusivity. When a pressurized permeable layer has a pressure diffusivity $\kappa_P = 1\,\text{m}^2/\text{s}$, a sudden pressure release following fracture diffuses approximately 4 cm in the first millisecond, 1 m in half a second and 1 km in about 6 days.

The maintenance of overpressure in deeply buried strata over geological time scales requires confinement by deep layers of very low permeability, possibly combined with the formation of new fluid by chemical reaction. The time scale for pressure diffusion varies, according to (3.106), as the square of the thickness of the confining overburden; for a layer of depth 10 km with $k \sim 10^{-19}\,\text{m}^2$, the relaxation time is some 2 million years, not long on a geological time scale.

3.6.2 *Pressure diffusion and de-gassing following seismic release*

A particularly noteworthy type of subterranean fluid conduit is the dilatational jog structure between fault segments produced during seismic events. These have been studied and described extensively by Sibson (1981, 1986, 1987). As illustrated in Figure 3.22, from Sibson (1987), they are localized zones of dilatation between laterally displaced fault segments that have undergone episodic relative displacement. They range in scale from centimeters to 1 km or so and are in the form of bundles of fissure veins or of bands of wall rock breccia (cemented rubble) that provide low-resistance pathways for vertical flow between the fault segments. Their textural characteristics testify to episodic efflux of hot (220–270 °C) hydrothermal fluids. Sibson estimates that as much as $10^7 \, m^3$ of fluid may have been ejected in a major seismic rupture, almost all of it through dilatational fault jogs. The strike-slip system itself is frequently almost free of mineralization, suggesting that the resistance to vertical flow in the planar fractures is much greater there than it is in the jogs themselves.

Fitzgerald and Woods (1994) and Tsypkin and Woods (2004, 2005) have discussed the dynamics of vapor fronts and precipitate formation, mainly in the contexts of geothermal circulation systems and reservoirs. Although the seismic events described by Sibson seem to have been much more violent and the conditions more extreme than those in established geothermal fields, the basic hydrogeology and geochemistry are similar. In the milliseconds after the initial rupture, the sudden reduction in pressure at the wall of the newly opened fissure in water-saturated rock over depths of up to 1 km and temperatures of 220–270 °C would cause an enormous transient pressure gradient in the interstitial fluid adjacent to the wall, with flash boiling of the pore water. The high-pressure, hot vapor ejecting into the fissure would produce, by its drag on the matrix, a tensile stress in the direction of flow. This would combine with the compressive lithostatic load to produce high shear stresses in the region near the fracture and a hydraulic implosion with rupture of the adjacent matrix into rubble and subsequent precipitation and mineral deposition. The outflow rate and its subsequent time history appear to be at least initially determined by the subterranean reservoir structure rather than the hydraulic resistance of the breccia-filled jog, which is generally small. Similar structures and distributions of mineralization, cited by Sibson (1987), have been identified in the Southern San Andreas fault system and in Waihi, New Zealand. Even though the hydrothermal fluids are hot during these events, possibly de-gassing and emerging from the surface as geysers, the dynamical effects of buoyancy on the flow are insignificant – the flow is certainly pressure-driven. Mineralization associated with the subsequent flow gradually reduces the porosity of the conduits within the jog, and the flow rate ultimately decreases.

3.6.3 Diffusion of pressure in a fracture–matrix medium

If the medium is extensively fractured, the diffusion of pressure following a sudden perturbation is much more rapid than in a uniform Darcy medium since, in essence, the characteristic block size replaces the domain size in determining the response time. The perturbation may be of seismic origin or it may be produced by a drilling bit as it breaks through an impermeable cap into a pressurized fluid reservoir. Interest in this application led to the analysis of pressure diffusion in a fracture–matrix medium pioneered by Barenblatt *et al.* (1960) using the concept of co-existing and interacting pressure fields defined by local averages over the resolution volume. It was they who developed many of the ideas used extensively in this book. Let p_M be the locally averaged pressure in the matrix blocks and p_F the average in the fracture network. Although on a microscopic scale, the pressure is continuous across the interfaces between fractures and matrix blocks, the local means p_M and p_F may differ. The internal pressure gradients in the matrix blocks are of order $(p_M - p_F)/l$, where l is the characteristic block size, and this drives fluid towards the fracture interface with transport velocity

$$u \sim \frac{k_M}{\mu} \frac{p_M - p_F}{l}$$

and flow divergence

$$Q = \nabla \cdot \mathbf{u} \sim u/l \sim \frac{k_M}{\mu} \frac{p_M - p_F}{l^2}. \tag{3.107}$$

This flow divergence represents a fluid sink (per unit volume) for the matrix blocks and a source for the fracture network. For the matrix blocks, the flow divergence is associated with decreasing fluid density and pressure as in (3.103):

$$\phi_M \frac{\partial \rho_M}{\partial t} = -\rho_0 \nabla \cdot \mathbf{u} = -\rho_0 Q$$

and, since again $\partial p = c_I^2 \partial \rho$,

$$\frac{\phi_M}{c_I^2} \frac{\partial p_M}{\partial t} + \frac{k_M}{\nu} \frac{(p_M - p_F)}{l^2} = 0$$

where ν is the kinematic viscosity of the interstitial fluid. This equation can be rearranged as

$$\frac{\partial p_M}{\partial t} + \frac{c_I^2 k_M}{\nu \phi_M l^2} p_M = \frac{c_I^2 k_M}{\nu \phi_M l^2} p_F. \tag{3.108}$$

The left-hand side of this equation specifies the pressure variations in the matrix blocks of size l when driven by the pressure variations in the fracture network (on the right). The fluctuations are damped with an attenuation time

$$t_M = \frac{\nu \phi_M}{c_I^2 k_M} l^2 = l^2 / \kappa_{P,M}, \tag{3.109}$$

where

$$\kappa_{P,M} = \frac{c_I^2 k_M}{\nu \phi_M}. \tag{3.110}$$

This can be interpreted as the pressure diffusivity in the fracture-matrix medium (c.f. equation (3.105)), but note that the permeability and porosity of the *matrix blocks* are involved. In terms of the dimensionless time, $\tau = t/t_M$, equation (3.108) is

$$\frac{\partial p_M}{\partial \tau} + p_M = p_F(\tau),$$

which is analogous to (3.91), with, here, the fluid pressure in the *matrix blocks* responding to variations in the fracture network pressure with a memory function that decays exponentially over the time scale t_M, above, rather than E^{-1}.

For the fracture network,

$$\phi_F \frac{\partial \rho_F}{\partial t} + \rho_0 \nabla \cdot \mathbf{u}_F = \rho_0 Q,$$

and since $\mathbf{u}_F = -(k_F/\mu)\nabla p_F$, $\mu = \rho_0 \nu$, and $\partial p = c_I^2 \partial \rho$, this becomes

$$\frac{\phi_F}{c_I^2} \frac{\partial p_F}{\partial t} - \frac{k_F}{\nu}\nabla^2 p_F = \frac{k_M}{\nu}\frac{(p_M - p_F)}{l^2} \tag{3.111}$$

$$\text{or} \quad \frac{\partial p_F}{\partial t} - \kappa_{P,F}\nabla^2 p_F = \frac{\phi_M}{\phi_F}\frac{p_M - p_F}{t_M}. \tag{3.112}$$

The left-hand side of this equation is again the classical heat conduction or diffusion operator with the fracture pressure diffusivity

$$\kappa_{P,F} = \frac{c_I^2 k_F}{\nu \phi_F} = \frac{k_F \phi_M}{k_M \phi_F}\kappa_{P,M} \tag{3.113}$$

and on the right is the pressure leakage from the matrix blocks, amplified by the ratio $(\phi_M/\phi_F) \sim 10^3$ because of the matrix/fracture volume ratio. By comparison with (3.110), it appears that the fracture pressure diffusivity is larger than the matrix pressure diffusivity by a factor of order 10^3 to 10^6, so that the distance that a pressure perturbation diffuses in a given time (i.e. $(\kappa t)^{1/2}$) through the fracture network is from 30 to 1000 as far as it does through a matrix block.

With these considerations in mind, one can infer the sequence of events that occurs when the confining layer of a saturated fracture–matrix region, under pressure from its overburden, is suddenly ruptured. The drop in fracture pressure diffuses rapidly throughout the region in accordance with (3.112), adjusting quickly to a quasi-steady state on a time scale $L^2/\kappa_{P,F}$, where L is the largest dimension of the region. The total fluid volume to be expressed from the fractures during this initial phase is relatively small because of the disparity in fracture and matrix porosities, but once the fracture pressure drops, fluid begins to be expelled from the matrix blocks over the longer time scale (3.109). Frequently, de-gassing accompanies the pressure drop with ejection of both steam and mineralizing fluids, as Sibson (1981, 1989) has described. Many interesting and important applications of fracture–matrix analysis to the extraction of hydrocarbons have been given by Barenblatt *et al.* (1990), which contains useful references to previous work.

4

Flows with buoyancy variations

4.1 The occurrence of thermally driven flows

A fundamental difference between constant density Darcy flows and those influenced by variations in buoyancy of the interstitial fluid is the absence of a uniqueness theorem in the latter situation. This is not just a matter of our being unable to prove such a theorem. In buoyancy-driven flows, different flow and temperature patterns can in fact occur in a given geometry and the same boundary conditions, and this raises new questions about the stability of flow patterns and the conditions under which one pattern can evolve into another. Internal circulating flows are prohibited in Darcy flows with uniform buoyancy, even though the permeability may vary arbitrarily, but now, flow rotation becomes of great interest, both in theory and in field observation.

Variations in buoyancy of interstitial fluids may be the result of variations in salinity, but are more commonly associated with temperature. The temperature variations that both drive subterranean flows, and are also modified by them, are generally the result of the geothermal heat flux from the earth's interior. The average upward heat flux in continental areas is about 2×10^{-6} cal cm^{-2} s^{-1}, a very small fraction of the mean incident solar energy flux. The mean temperature gradient in the earth's upper crust is some 2–3 °C per 100 m, but this gradient and the local heat flux are concentrated and augmented in volcanic and hydrothermal areas by convective heat transfer associated with bodily fluid movements. Geothermal areas in Iceland, the western United States, in Italy, New Zealand and elsewhere have been studied extensively, although the detailed structure of these regions and their internal plumbing is still to some extent conjectural.

Pure thermally driven flows are generally to be found only when the hydraulic forcing is absent, in geothermal regions, in submerged banks, beds, or continental shelf regions or in isolated, totally confined permeable strata. The total pressure at the bed of the sea or a lake is very nearly equal to the hydrostatic head of the

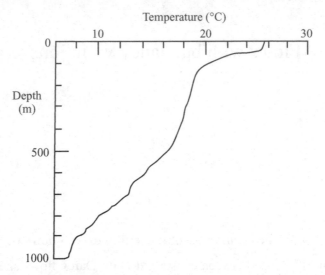

Figure 4.1. An oceanic temperature profile, measured by Hogg, Katz and Sanford (1978) off Bermuda in October 1975, is typical of those found in the top 1000 m in moderate latitudes. Note the shallow thermocline at 50–100 m depth. The ocean temperature almost always decreases with depth, whereas the temperature in the crust increases with depth at an average rate of about 30 °C/km.

water above, so that the reduced pressure at the sediment–water interface is usually very small. It is the *reduced pressure gradient* that supports hydraulically driven flow, and if a permeable region is isolated by impermeable rock on all sides with no fluid entering or leaving it, any internal motion must be driven thermally. In any of these regions, if for any reason the isotherms are not horizontal, interstitial fluid will circulate within the permeable domain, as specified by the rotation vector equation (4.3), below.

In marine sediments, the internal isotherms will slope when the sediment–water interface slopes or when the geothermal heat flux varies or when both occur, and the consequences can be discerned in ancient marine sediments, now uplifted and long dormant. At the upper surface of a submerged bank or coral reef, the rapid heat transfer from ocean mixing ensures that the interfacial temperature of the sediment is close to that of the water above. In tropical or temperate latitudes, the near-surface sea-water mixed by the wind has temperatures of approximately 20–25 °C, varying somewhat with the season. Below this mixed layer, the temperature decreases rapidly with depth in the thermocline (Figure 4.1), to values of only a few degrees Celsius below depths of 1000 m or so. Here, the temperature remains nearly constant at all latitudes throughout the year. In high latitudes, the oceanic water column is cold throughout its depth; in fact, the cold, deep water in all the world's oceans is maintained by the subsidence of Arctic and Antarctic Ocean water from the surface to great depths, the gradual meander south and north, and the eventual upwelling and incorporation into the oceanic surface layers. The geothermal heat flux through the ocean floor has a generally imperceptible influence on this oceanic temperature distribution except locally in the vicinity of thermally active vents. In most of these particular areas, hot plumes from deep oceanic vents rise and mix quite rapidly with the surrounding water, though in a few regions devoid of deep ocean circulation (the Red Sea being a notable example) hot, dense brine pools, fed by vents, fill the greatest depths. There is good geological evidence that massive mineralogical changes have been produced in relatively shallow marine sediments by long-lived and extensive convection patterns associated with tectonic heat anomalies, as discussed by Bosellini and Rossi (1974), Gaetani *et al.* (1981), Wilson *et al.* (1990) and others.

Relatively weak internal circulations may occur in coral reefs or islands that are frequently composed of high-permeability reef limestone, interlayered with approximately horizontal, lithified lagoon muds and silts of much lower permeability. The internal temperature gradients there are much smaller than they are in geothermal areas. The layered structure of the medium offers greater resistance to interior vertical flow than it does to horizontal flow inward from the surrounding escarpments. Unless the structure contains fractures or internal conduits, the convective circulation is dependent on the capacity of the thermally induced buoyancy variations to drive vertical motion through the more resistive, less-permeable layers.

4.2 Buoyancy and the rotation vector

The basic dynamical statement, Darcy's equation (2.24), expresses the transport velocity in terms of the distributions of reduced pressure and interstitial fluid buoyancy

$$\mathbf{u} = \frac{k}{\nu}(-\nabla(p/\rho_0) + b\mathbf{l}) \quad \text{where the buoyancy} \quad b = g\left(\frac{\rho_0 - \rho}{\rho_0}\right), \qquad (4.1)$$

ν is the kinematic viscosity, g represents the gravitational acceleration and ρ_0 is the reference density of the fluid, often the mean or the value beyond the region of active flow. The buoyancy force b acts vertically upward or down. The buoyancy itself is also a field variable, a function of temperature T and salinity S through an equation of state such as (2.63):

$$\rho = \rho_0(1 - \alpha T + \beta S), \quad \text{so that} \quad b = g(\alpha T - \beta S), \qquad (4.2)$$

though a nonlinear form may be more appropriate at low solute temperatures. In the case of water and most dilute solutions, the coefficient of thermal expansion α

is negative near 0 °C, about 9×10^{-5} °C^{-1} in a small temperature range around 10 °C, and increasing to 3.5×10^{-4} °C^{-1} near 35 °C. When the salinity S is measured in grams of solute in each kilogram of solution (parts per thousand), the coefficient $\beta \sim 0.7$ (Weast, 1972). The separate distributions of temperature and salinity are coupled to the velocity field through the conservation equations (2.48) for heat and (2.52) for salt. Chemically passive dissolved salts move through the medium at the interstitial fluid velocity and can be transported much more rapidly than heat through a classical Darcy "sandbank" medium as described in Sections 2.8 and 2.10, and even more so in a fracture matrix medium (Section 3.5). This differential effect gives rise to a whole class of so-called double-diffusive instabilities and other phenomena, described extensively by Turner (1973) and encountered in permeable medium flows as well as in bulk fluids.

The general characteristics of buoyancy-driven flows differ in some important respects from those of pure Darcy–Laplace flows described in the last section. Buoyancy-driven flows are almost always rotational and the rotation vector of the transport velocity field (3.5), i.e.

$$\Omega = (\Omega_X, \Omega_Y, \Omega_Z) = \nabla \times \mathbf{u} = \mathrm{curl}\,\mathbf{u}$$

plays an important role in their dynamical behavior. The curl of the Darcy equation (4.1) is

$$\Omega = \nabla \times \mathbf{u} = \frac{k}{\nu}\{\nabla \times (b\mathbf{l})\}, \qquad (4.3)$$

where, again, $\mathbf{l}$ is a unit vector, vertically upward. The pressure term has disappeared since $curl(grad)$ is identically zero. In terms of Cartesian coordinates (x, y, z) with the z-axis vertically upward, and velocity components (u, v, w)

$$\Omega_X = \frac{\partial w}{\partial y} - \frac{\partial v}{\partial z} = \frac{k}{\nu}\frac{\partial b}{\partial y},$$

$$\Omega_Y = \frac{\partial u}{\partial z} - \frac{\partial w}{\partial x} = -\frac{k}{\nu}\frac{\partial b}{\partial x},$$

$$\Omega_Z = 0. \qquad (4.4)$$

It is evident that the component of the rotation vector in each horizontal direction is proportional to the slope of the lines of constant density or buoyancy (the isopycnals) in the orthogonal horizontal direction. From the last equation in the set, the vertical component of the rotation vanishes. This fact implies from (4.4) that even in flows influenced by buoyancy, the circulation around every closed circuit in a *horizontal* plane must vanish. In plan form, there can be no closed streamlines or flow paths – the horizontal part of the flow pattern is irrotational. Rising plumes in Darcy flow do not swirl.

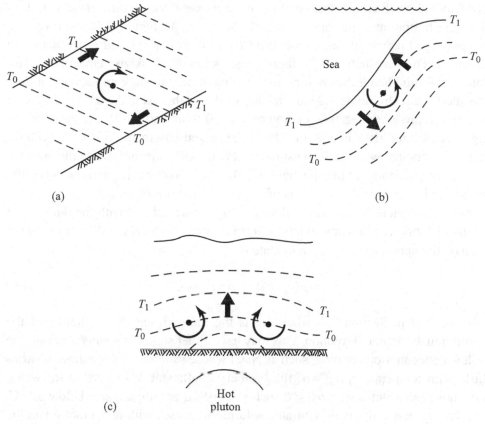

Figure 4.2. Cartoons of the flows induced in permeable media by horizontal temperature gradients. In each diagram, $T_0 > T_1$. When the isotherms slope downward to the right, the rotation vector is clockwise and when they slope upward, the rotation is counter-clockwise.

Some simple two-dimensional cartoons of buoyancy-driven flows are shown in Figure 4.2, in which the variations in buoyancy are the result of temperature variations in the x–z plane. The rotational flow in this plane is given by (4.4) as

$$\Omega_Y = -\frac{k}{\nu}\frac{\partial b}{\partial x} = -\frac{\alpha g k}{\nu}\frac{\partial T}{\partial x}, \qquad (4.5)$$

with use of the linear approximation to the equation of state. The isotherms are dashed and in each sketch, $T_0 > T_1$. In Figure 4.2a the permeable region is bounded by sloping impermeable layers and the isotherms slope downward to the right. Fluid near the top of the porous layer is hotter and more buoyant than fluid in the interior at the same level, so that $\partial T/\partial x < 0$. It therefore moves upward and to the right *relative to* the fluid near the bottom. Superimposed on this may be an arbitrary, pressure-driven flow, uniform across the width of the porous layer, but even in the combined flow, fluid near the top of the layer moves *faster* to the right than it does near the bottom and the rotation is still clockwise. In Figure 4.2b, representing a submerged permeable escarpment, $\partial T / \partial x > 0$ so that the sense of rotation is reversed. This distribution of isotherms results in anti-clockwise circulation, with interstitial fluid being drawn into the escarpment near its base and leaving it near the shelf break. In Figure 4.2c, the raising of the isotherms in the medium above a hot pluton produces negative rotation on the left flank and positive rotation on the right as the fluid rises in a plume above the heated basement region. The central hottest interstitial water is the most buoyant and moves upward, while the outward buoyancy gradients of opposite sign on either side produce the shear between the plume and the ambient fluid that is of opposite sign on each side.

When variations in interstitial fluid density are associated with the dissolution of major fabric constituents and the solution is close to local equilibrium with the matrix, the appropriate equation of state is

$$\rho = \rho(c_S(T)) \approx \rho_0(1 - \alpha_S T), \tag{4.6}$$

as described in Section 2.9, where α_S is the thermal density coefficient of the continuously saturated solution. This may have either sign. For solutes that saturate at low concentrations or those such as NaCl whose saturation concentration varies little with temperature, $\alpha_S \approx \alpha$, the ordinary coefficient of expansion for water. For those few solutes with $\alpha_S < 0$ such as Na_2SO_4 at temperatures below 32 °C, the density of a continually saturated solution increases with temperature and the senses of rotation shown in Figure 4.2 are *reversed*.

4.3 General properties of buoyancy-driven flows

In general hydro-geological flows, the temperature and salinity are field variables that separately (i) satisfy conservation equations (2.48) and (2.52), (ii) enter into the force balance through the buoyancy term in Darcy's equation (2.20), (2.24) or its counterparts in fracture–matrix media and (iii) are interconnected in an equation of state such as (4.2). For convenience, these are reproduced below.

Heat conservation, first law of thermodynamics:

$$M\frac{\partial T}{\partial t} + \mathbf{u} \cdot \nabla T = \kappa \nabla^2 T + Q. \tag{2.48, 4.7}$$

Salt conservation:

$$\frac{\partial S}{\partial t} + \bar{\mathbf{v}} \cdot \nabla S = D\nabla^2 S + Q_S. \tag{2.52, 4.8}$$

Force balance in Darcy flow:

$$\mathbf{u} = \frac{k}{\nu}\left(-\nabla(p/\rho_0) + b\mathbf{l}\right). \tag{2.24, 4.9}$$

The incompressible form of mass conservation:

$$\nabla \cdot \mathbf{u} = 0. \tag{2.8, 4.10}$$

Linear equation of state:

$$\rho = \rho_0(1 - \alpha T + \beta S), \tag{4.11}$$

where, in (4.9), the buoyancy $b = g(\rho_0 - \rho)/\rho_0$.

4.3.1 Heat advection versus matrix diffusion: the Peclet number

In many situations, particularly with the dilute solutions characteristic of natural flows, internal heat sources from heats of reaction and salt sources from precipitation or dissolution are quite negligible compared to the advective and diffusive flux divergences of heat (for example, in the flow-controlled deposition of calcite, the ratio is of order 10^{-6}; see, for example, Bathurst, 1975, p. 258). When also the distributions of temperature and salinity are close to steady state, the first and last terms of equations (4.7) and (4.8) can be dropped, and finally, when the buoyancy is the result of temperature variations alone, the set of balances reduces to

$$\mathbf{u} \cdot \nabla T = \kappa \nabla^2 T$$
$$\mathbf{u} - \frac{k}{\nu}\{-\nabla(p/\rho_0) + g\alpha T\mathbf{l}\} \tag{4.12}$$
$$\nabla \cdot \mathbf{u} = 0.$$

Recall that κ, the thermal diffusivity, is the matrix thermal conductivity divided by ρC_F, and the unit vector $\mathbf{l}$ is vertically upward.

These coupled equations for the transport velocity $\mathbf{u}$, reduced pressure p and temperature field T are slightly nonlinear through the advection term $\mathbf{u} \cdot \nabla T$. The pressure can be eliminated from them by cross-differentiation as in the rotation vector expression (4.3), or expressed directly in terms of the buoyancy field by taking the divergence of the second of (4.12):

$$\nabla^2 p = \rho_0 g\alpha \frac{\partial T}{\partial z}. \tag{4.13}$$

The anatomy of the first equation of the set (4.12) is of interest. In this heat balance, the advection term on the left expresses the heat flux divergence as the product of the fluid transport speed and the temperature gradient in the direction of flow, and is of order $u\Delta T/l$, where ΔT is the characteristic magnitude of the

temperature variations *along the streamlines* and l is the distance over which
this variation occurs. Individual terms in the isotropic Laplacian diffusive term
$\nabla^2 T$ are of order $\kappa \Delta T / h^2$, where h is the smallest geometrical length scale in the
flow domain, usually the flow layer thickness. The advective heat transfer by the
interstitial fluid dominates thermal diffusion through the matrix when $u \Delta T / l \gg$
$\kappa \Delta T / h^2$, i.e. when

$$Pe = \frac{uh}{\kappa} \gg \frac{l}{h}. \tag{4.14}$$

The dimensionless ratio $Pe = uh/\kappa$, known as the Peclet number, expresses the
ratio of the advective to diffusive heat transport, and l/h is the aspect ratio of the
domain. When the Peclet number is large enough to satisfy the condition (4.14), a
situation generally involving highly permeable aquifers that are not too long and
having a large flow rate, heat diffusion is insignificant and the first equation in
(4.12) might perhaps be approximated by

$$\mathbf{u} \cdot \nabla T = 0, \tag{4.15}$$

which asserts that the temperature is constant along the lines of flow.

But this now poses a contradiction. In a closed region or along an impermeable
boundary, the fluid flow follows the boundary, yet heat must be conducted *across*
the boundary since the motion is generated thermally. In short, (4.15) does not
allow us to satisfy the thermal boundary conditions. This kind of problem is
characteristic of equations with the structure of (4.7) viewed as an equation for the
temperature field T in which a small parameter κ (or in dimensionless terms Pe^{-1})
multiplies the highest derivative term in the equation ($\nabla^2 T$ is second order; ∇T is
first order). Equations of this type are called singular perturbation equations. If the
term of highest order is dropped, the order of the differential equation is reduced
and fewer boundary conditions can be satisfied. The solution to the dilemma lies
in the realization that the general scaling cannot be true everywhere – that there
must be regions of the flow, involving plumes or thermal boundary layers, which
are sufficiently thin compared with the layer thickness that the second-derivative,
thermal-diffusion term is *locally* comparable with the advection term, even though
over most of the fluid, it does indeed remain negligible. In these large Peclet
number flows, then, there are different domains – an interior region away from
boundary surfaces or plumes in which (4.15) is valid and the temperature does
remain very nearly constant along the lines of flow, together with relatively thin
boundary regions whose width self-adjusts to produce a balance between convective
and diffusive effects in the heat equation. The solution to the flow as a whole
involves matching these various regions to ensure that both the temperature and
velocity distributions join smoothly from one region to another, a procedure that is

exemplified later in this chapter. The temperature and flow fields must therefore be solved simultaneously rather than sequentially before patterns of fabric alteration can be inferred.

At the other extreme, when the Peclet number is small, the advective effects of heat transport on the temperature are negligible, and so within an aquifer the temperature is determined solely by thermal conduction, as it may also be in the underlying and overlying aquistads:

$$\nabla^2 T = 0, \qquad \text{when } Pe \ll (l/h). \tag{4.16}$$

In either case, then, the temperature distribution can be found fairly simply; it is only when advection and diffusion of heat are comparable that the groups of terms on the two sides of the first equation in the set (4.12) are of the same order and a detailed numerical or analytic solution may be necessary.

4.3.2 Thermally driven flows: the Rayleigh number

Just as hydraulic pressure gradients are generally dominant in the flow through aquifers, so in isolated, fluid-saturated regions or in submarine sediments and fracture regions where there are no hydraulically imposed pressure gradients, flow will be induced unless the isopycnals (surfaces of constant density) happen to be precisely horizontal. In a few cases, horizontal density variations can be the result of horizontal variations in salinity, as in salt wedges, as described later. Horizontal variations in salinity also occur flanking salt domes, as Evans and Nunn (1989) have pointed out, and these may induce salinity-driven convective motion. The interplay between salinity and temperature variations in driving a flow field can be quite intricate, in some circumstances destabilizing a basic state that would appear to be stable and in other circumstances stabilizing one that seems unstable. These considerations will be deferred until later in this chapter. For the present scaling purposes, let us consider flows that are thermally driven, those in which the isotherms are not horizontal. The flow and temperature fields are more intimately connected than they are in hydraulically driven flows, but again, scaling considerations allow significant simplifications in many cases.

When the variations in interstitial fluid buoyancy are the result of variations in temperature, the three Cartesian components of the rotation vector (4.4) are

$$
\begin{aligned}
\Omega_X &= \frac{\partial w}{\partial y} - \frac{\partial v}{\partial z} = \frac{gk\alpha}{\nu} \frac{\partial T}{\partial y}, \\
\Omega_Y &= \frac{\partial u}{\partial z} - \frac{\partial w}{\partial x} = -\frac{gk\alpha}{\nu} \frac{\partial T}{\partial x}, \\
\Omega_Z &= 0.
\end{aligned}
\tag{4.17}
$$

When thermally driven flows occupy an extensive region, i.e. when the regions of significant horizontal temperature gradient driving the flow are of larger extent horizontally than vertically, the flow may break up into a series of cells with roughly comparable vertical and horizontal scales. Because of the form of the incompressibility condition, the scale of horizontal and vertical transport velocities are also generally comparable, $u \sim w$. From (4.17), we note that the induced transport velocity gradients are proportional to the *horizontal* temperature gradient, so that the magnitude of the rotation vector is

$$|\Omega| \sim \frac{u}{h} \sim \frac{gk\alpha}{\nu}\frac{\Delta T}{l},$$

where ΔT is temperature difference over the horizontal distance l. Consequently, the representative magnitude of the flow velocity is

$$u \sim \frac{kg\alpha(\Delta T)}{\nu}\left(\frac{h}{l}\right) = K\alpha(\Delta T)\left(\frac{h}{l}\right), \tag{4.18}$$

where $K = kg/\nu$ is the hydraulic conductivity. In a compact flow domain, with $l \sim h$ and $u \sim w$, this reduces to

$$u \sim \frac{kg\alpha(\Delta T)}{\nu} = K\alpha(\Delta T). \tag{4.19}$$

These are very useful general velocity scales for steady, thermally driven flow. In the numerator, one observes the physical quantities that promote the flow, the permeability k and the buoyancy $g\alpha(\Delta T)$; in the denominator is the fluid viscosity that impedes it. One also notes from the second equality that the transport velocity is given by the hydraulic conductivity (which has the physical dimensions of velocity) times the fractional density difference involved. In such a flow, the Peclet number $Pe = uh/\kappa$ (which expresses the ratio of advective to diffusive heat transport) assumes the form

$$Pe = \frac{uh}{\kappa} = \frac{gkh\alpha(\Delta T)}{\nu\kappa} = Ra, \text{ say,} \tag{4.20}$$

where κ is the thermal diffusivity. The combination

$$Ra = \frac{gkh\alpha(\Delta T)}{\nu\kappa} = \frac{Kh\alpha(\Delta T)}{\kappa} \tag{4.21a}$$

is called the Rayleigh–Darcy number or simply the Rayleigh number if the context is clear. An alternative form in terms of the vertical temperature gradient is

$$Ra = \frac{gkh^2\alpha}{\nu\kappa}\left(\frac{\partial T}{\partial z}\right). \tag{4.21b}$$

It is only in compact thermally driven flows that the numerical values of the Peclet and Rayleigh numbers are equal. The Rayleigh number appears later in this chapter as an important parameter in questions of the structure and stability of thermally driven flows, but here, because of its association (4.20) with the Peclet number, it is also an index of the relative importance of heat convection and thermal diffusion. In small Rayleigh number flows, the temperature distribution (and the consequent velocity field) is determined primarily by thermal conduction through the matrix. When $Ra = Pe \gg 1$, the flow and temperature fields are convectively coupled, as shown above.

4.4 Steady low Rayleigh number circulations

4.4.1 Slope convection with large aspect ratio l/h

In a submerged permeable bank, the internal isotherms associated with the geothermal heat flux below the interface are generally parallel to the interface, being horizontal when the sea-bed is flat and sloping upward inside an escarpment. The slope of the isotherms generates a generally weak rotational flow, drawing fluid in near the base of the escarpment and releasing it near the top, as illustrated in Figure 4.2b. In this kind of flow, dissolution and dolomitization reactions may occur near the base as the interstitial fluid is drawn into regions of higher internal temperatures, and reversing as the flow returns to the cooler interfacial region above.

The general flow magnitude can be estimated without calculation from the thermally induced velocity scale (4.18), in the form

$$u \sim K\alpha(\Delta T)\left(\frac{h}{l}\right) \sim \frac{\kappa\, Ra}{l}. \tag{4.22}$$

Recall that ΔT is the temperature difference within the bank over the horizontal distance l and that h/l ($\ll 1$) is the slope of the internal isotherms. Far from the edge of the bank, the isotherms are horizontal and no flow is induced there. In a two-dimensional bank such as illustrated in Figure 4.3, the total volume flux per unit length is proportional to

$$q \approx uh = Kh^2\alpha(\partial T/\partial z) = Ra\frac{\kappa h}{l},$$

but the numerical coefficients found from the calculation are small. A convenient flow magnitude for comparison might be that in a surface aquifer in a temperate climate with mean rainfall rate of 2 m/yr or somewhat more, with infiltration

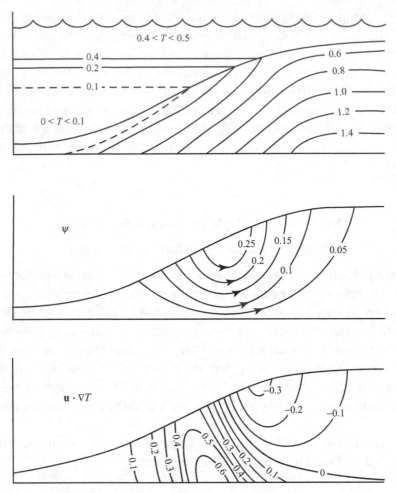

Figure 4.3. Calculated isotherms, streamlines and distributions of the alteration index $\mathbf{u} \cdot \nabla T$ in a submerged escarpment with a "tanh" profile and height h above an impermeable basement. The surrounding water contains a shallow thermocline that abuts the escarpment. The vertical geothermal gradient G deep inside the bank is constant and the temperature of the water–sediment boundary is that of the adjacent water. The temperature scale is Gh, that of the stream function is $Ra(\kappa h/l)$ and of $\mathbf{u} \cdot \nabla T$ is $Ra(G\kappa h/l^2)$.

(transport velocity downward) perhaps half of this. This enters the aquifer across its upper surface and the total flux along the aquifer per unit width near the discharge region is equal to the infiltration rate times the aquifer length, which may be 10 km or 10^4 m . In a surface aquifer of moderate size, then, a representative volume flux per unit width is of the order of 10^4 m²/yr. In a submerged bank like that shown in Figure 4.3, the value of the stream function (the magnitude of the flux per unit

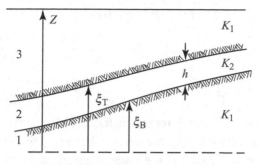

Figure 4.4. Definition sketch.

width) is of order 0.1 in units of $Ra(\kappa h/l)$. The low Rayleigh number calculation is for $Ra(h/l) \ll 1$, or about 0.1 at most. The thermal diffusivity κ of saturated sandy material is about $10^{-7}\,\text{m}^2/\text{s} \sim 3\,\text{m}^2/\text{yr}$, so that the thermally induced volume fluxes are of order $3 \times 10^{-2}\,\text{m}^2/\text{yr}$, which is very small. Unless there are internal conduits that concentrate the flow, mineral alteration resulting from slope convection is likely to be diffuse, and possibly of less mineralogical significance than other, more intense and more localized flow processes.

4.4.2 Circulation in isolated, sloping permeable strata

Convective circulations driven by the geothermal temperature gradient may also occur in isolated permeable strata, as shown qualitatively by Wood and Hewett (1982) and quantitatively by Davis *et al.* (1985). These can redistribute the pattern of mineral deposition even in an isolated closed region. Consider a gently sloping layer of saturated permeable sandstone lying between impermeable layers of shale (Figure 4.4). Circulating flow in the permeable layer is again produced by sloping isotherms, resulting from differences between thermal conductivity of the matrix and the layer, even when the ambient geothermal gradient is vertical. The steady temperature distributions in the shale and, at low Rayleigh numbers, in the saturated layer also, are determined by Laplace's equation. With the neglect of terms of relative magnitude $(h/l)^2 \ll 1$, this reduces to

$$\partial^2 T / \partial z^2 = 0 \qquad (4.23)$$

in each region. Note that in this example, l represents the length of a horizontal traverse of the layer, so that h/l is the characteristic dip of the layer, which we assume to be small. Across the upper and lower interfaces of the permeable region, the temperatures and normal heat fluxes must be continuous, so that again neglecting slope terms of order $(h/l)^2$, we have

$$\text{at} \quad z = \zeta_B, \qquad T_1 = T_2, \qquad \kappa_1 \frac{\partial T_1}{\partial z} = \kappa_2 \frac{\partial T_2}{\partial z},$$

$$\text{at} \quad z = \zeta_T \qquad T_2 = T_3, \qquad \kappa_2 \frac{\partial T_2}{\partial z} = \kappa_3 \frac{\partial T_3}{\partial z}, \qquad (4.24)$$

where the subscripts 1, 2, and 3 represent the regions below, in, and above the permeable layer, respectively, and the κs are the respective matrix thermal conductivities as defined by (2.48). If the upper impermeable region is of the same mineralogy as the lower one, $\kappa_1 = \kappa_3$. Below the layer, the temperature distribution is geothermal $\partial T / \partial z = -G$, which will be supposed horizontally uniform. At the overlying land surface $z = Z$, $T = 0$, say.

From (4.23), the temperature distribution in each region must be of the form

$$T = A(x) + zB(x), \qquad (4.25)$$

where A and B may be slowly varying functions of horizontal position. In the lowest region, 1, the vertical temperature gradient is the undisturbed geothermal, so that $B_1 = -G$ and

$$T_1 = A_1(x) - zG.$$

After substitution of expressions such as (4.25) into the conditions (4.24) and with the condition that $T_3(z) = 0$, we have in all five algebraic relations for the five other coefficient functions A_1, A_2, A_3, B_2, B_3. After a little algebra, the solution for the temperature distribution in the upper, middle and lower regions is found to be

$$\begin{aligned}
T_1 &= G\{(z_0 - z) - \{(1 - \kappa_1/\kappa_2)(\zeta_T(x) - \zeta_B(x))\}\}, \\
T_2 &= G\{(z_0 - (\kappa_1/\kappa_2)z - \{(1 - \kappa_1/\kappa_2)\zeta_T(x)\}\}, \qquad (4.26) \\
T_3 &= G(z_0 - z),
\end{aligned}$$

as given by Davis *et al.* (1985).

Of greatest interest is the temperature distribution in the permeable layer, 2. If the height of its upper surface varies in the horizontal direction, this permeable region has a horizontal temperature gradient

$$\nabla_H T = -G\{1 - \kappa_1/\kappa_2\}\nabla_H \zeta_T \qquad (4.27)$$

that is proportional to the slope of the top of the layer and the ratio of thermal diffusivities. In Figure 4.5, this slope is positive and if the layer has a higher thermal diffusivity than its surroundings, $\kappa_2/\kappa_1 > 1$, and, from (4.27), $\nabla_H T$ is negative (temperature decreasing along a horizontal to the right). The rotation vector is then positive into the paper and the circulation is clockwise. When $\kappa_2/\kappa_1 < 1$, the sense of the circulation is reversed.

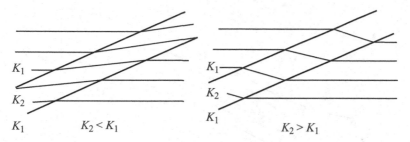

Figure 4.5. Inclined isotherms in a sloping layer with a thermal conductivity different from that in the matrix above and below and a uniform vertical heat flux.

The potential importance of this circulation is that in a chemically and hydro dynamically closed system, it provides a mechanism for the gradual increase in concentration of mineral deposits from solution in some places and depletion in others. In two dimensions, where the permeable layer is a sheet folded to form an anticline, the circulation can be expressed in terms of the stream function ψ, and from the rotation vector relation (4.4),

$$\frac{\partial^2 \psi}{\partial z^2} = -\frac{gk\alpha}{\nu} \nabla_H T = -K\alpha \nabla_H T \tag{4.28}$$

in terms of the hydraulic conductivity K and the thermal expansion coefficient α. Since both upper and lower interfaces are streamlines,

$$\psi = \frac{1}{2} K\alpha G \left(1 - \kappa_1/\kappa_2\right)\left(\partial \zeta_T/\partial x\right)\left(z - \zeta_B\right)\left(\zeta_T - z\right). \tag{4.29}$$

Streamline patterns calculated by Davis *et al.* from this expression, and distributions of the alteration index $\mathbf{u} \cdot \nabla T$ (c.f. Section 5.5) for this flow, are shown in Figure 4.6 for a ratio $\kappa_2/\kappa_1 = 1.25$. Note that the stream function is proportional to the vertical temperature gradient, so that the rates of mineral alteration are at least quadratic in this quantity. If the equilibrium concentration of solute increases with temperature, mineral deposition tends to occur in regions where the fluid is moving to lower temperatures, that is, along the upper interface where the flow is down the temperature gradient. Dissolution occurs in regions where fluid is moving to higher temperatures, along the lower interface. The most active regions of diagenesis are on the flanks, with regions of opposite sign occupying the upper and lower parts of the permeable layer. Quartz, for example, tends to precipitate along the upper flanks while dissolving along the lower slopes. When $\kappa_2 < \kappa_1$, the circulations and patterns of precipitation/dissolution are reversed.

The processes of dissolution, transport and precipitation will presumably continue until the dissolving mineral is locally exhausted or some other geological

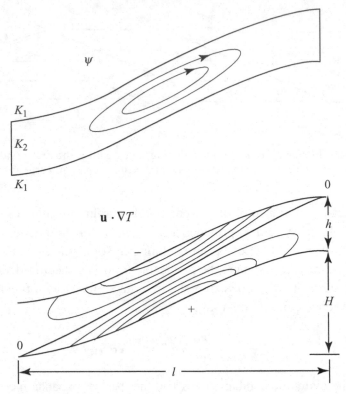

Figure 4.6. Streamlines and (below) contours of the alteration index $\mathbf{u} \cdot \nabla T$, in a permeable anticline when $\kappa_2 = 1.25\kappa_1$ (Davis *et al.*, 1985). When $\kappa_1 > \kappa_2$, the sense of the circulation is reversed and the regions of positive and negative $\mathbf{u} \cdot \nabla T$ are interchanged.

change occurs. Within the overall pattern, flow and reaction sites are frequently concentrated in smaller scale cracks or fractures.

4.4.3   Compact layered platforms and reefs at low Rayleigh numbers

In the foregoing examples of convectively driven flow at large aspect ratio l/h, the large horizontal extent allowed fluid to traverse the relatively short vertical distances more freely than over the large horizontal distances. In that case, the circulating convection velocities were found to be proportional to the horizontal *gradient* or differences of temperature but in compact flows, a different balance is found. Carbonate reefs or sedimentary banks are frequently strongly horizontally layered, so that their permeability k_V to vertical flow is considerably less than k_H for horizontal flow, whereas their horizontal extent may be only a moderate multiple of their thickness. Interstitial fluid can more freely move laterally from the ambient sea into the platform or over relatively short horizontal distances, but the convective circulation in the interior is limited by the resistance to the more central rise of buoyant fluid across the less permeable layers. When the permeability anisotropy is large, then, one would anticipate that at any *Ra* the vertical velocity in an interior flow (outside any boundary layers) is proportional to the *temperature itself* (above ambient), not its gradient.

This expectation is supported by an examination of the relation between the rotation vector and the buoyancy field in a medium with fine-scale, strong laminations in permeability. The extension of (4.4) to this situation is

$$\frac{\partial(w/k_V)}{\partial y} - \frac{\partial(v/k_H)}{\partial z} = v^{-1}\frac{\partial b}{\partial y},$$
$$\frac{\partial(u/k_H)}{\partial z} - \frac{\partial(w/k_V)}{\partial x} = -v^{-1}\frac{\partial b}{\partial x}. \tag{4.30}$$

In this pair of equations, v is the fluid viscosity. The horizontal and vertical dimensions of the flow domain (x, y, z) are generally comparable as are the velocity components (u, v, w), but $k_H \gg k_V$, so that the terms involving k_H on the left are very much smaller than the others. When they are ignored, the set (4.30) reduces to

$$\nabla_H(w/k_V) = \frac{g\alpha}{v}\nabla_H T. \tag{4.31}$$

Thus

$$w(\mathbf{x}, z) = K_V\alpha T(\mathbf{x}, z) + f(z), \tag{4.32}$$

where K_V is the hydraulic conductivity involving the permeability for vertical flow and $f(z)$ is an arbitrary function of z, which, physically, would represent the contribution from any hydraulic forcing. If $w = 0$ when $T = 0$ at, say, the edges of a bank, then $f(z) = 0$. It is evident from the Darcy balance (4.32) that in these compact interior regions with fine-scale, strong horizontal laminations, the local fluid buoyancy balances the local resistance to vertical flow. Note that this statement depends only on the geometry of the flow domain and the inequality $k_H \gg k_V$ and so is independent of the Rayleigh number

When *Ra* is small, however, the vertical temperature gradient is unaffected by the flow and

$$\frac{\partial w}{\partial z} = -\nabla_H \cdot \mathbf{u} = k_V\alpha\frac{\partial T}{\partial z}, \tag{4.33}$$

where $\partial T/\partial z = -G$. Thus $\partial w/\partial z < 0$, the vertical flow is convergent and $\nabla_H \cdot \mathbf{u} > 0$, the horizontal flow is divergent; in the interior regions of compact flows with fine-scale, strong laminations, when *Ra* and $(h/l)^2(k_H/k_V) \ll 1$, the rising vertical velocity always decreases upwards and the horizontal flow always

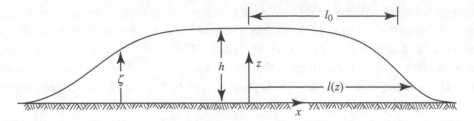

Figure 4.7. A compact platform; definition sketch.

diverges. These conclusions are qualitatively different from those found earlier in this section for convective flows in domains with high aspect ratio l/h.

Suppose the submerged permeable reef or bank is on a horizontal, relatively impermeable basement and that the configuration of the sediment–water interface is specified by $\zeta(\mathbf{x})$ (see Figure 4.7). If the ratio of width to thickness is not too small (i.e. not less than 5 or so) and the Rayleigh number is small, the internal temperature distribution driving the convection is still given to sufficient accuracy by the water temperature at the upper surface (0, say) plus the increase with depth from the geothermal gradient: $T(\mathbf{x}, z) = G(\zeta(\mathbf{x}) - z)$.

The analysis of these compact flows is most revealing in terms of the reduced pressure, which vanishes at the upper surface $z = \zeta(\mathbf{x})$ of the region. The horizontal and vertical force balances expressed by the Darcy equations are

$$\nabla_{\mathrm{H}} p = -\frac{\mu}{k_{\mathrm{H}}} \mathbf{u},$$

$$\frac{\partial p}{\partial z} = -\frac{\mu}{k_{\mathrm{V}}} w + \rho_0 g \alpha T,$$

whence, by use of the incompressibility condition $\nabla_{\mathrm{H}} \cdot \mathbf{u} + \partial w / \partial z = 0$, we have

$$k_{\mathrm{H}} \nabla_{\mathrm{H}}^2 p + k_{\mathrm{V}} \frac{\partial^2 p}{\partial z^2} = -\rho_0 k_{\mathrm{V}} g \alpha G, \tag{4.34}$$

where ∇_{H} and ∇_{H}^2 represent the two-dimensional, horizontal divergence and Laplacian operators. It is convenient to use dimensionless variables with l and h as vertical and horizontal length scales so that $(X, Y, Z) = (x/l, y/l, z/h)$. The scale for the internal reduced pressure is chosen as

$$p_{\mathrm{S}} = \rho_0 g \alpha G k_{\mathrm{V}} l^2 / k_{\mathrm{H}}, \tag{4.35}$$

which involves the vertical buoyancy gradient $g \alpha G$, the size of the region, l^2, and the permeability ratio $k_{\mathrm{V}}/k_{\mathrm{H}}$. In terms of the dimensionless pressure $P = p/p_{\mathrm{S}}$, equation (4.34) becomes

$$\frac{\partial^2 P}{\partial X^2} + \frac{\partial^2 P}{\partial Y^2} + \left\{ \frac{l^2}{h^2} \frac{k_V}{k_H} \right\} \frac{\partial^2 P}{\partial Z^2} = -1, \tag{4.36}$$

which is subject to the boundary conditions (i) on the interface between the bank and the water above, the reduced pressure p vanishes so that $P = 0$ on $z = \zeta(\mathbf{x})$, and (ii) that there is no flow across the lower boundary so that $w = 0$ on $z = 0$. On this surface, the temperature relative to that of the ambient water is $G\zeta(\mathbf{x})$ where G is the geothermal gradient, so that from the vertical Darcy balance above, this boundary condition is

$$\frac{\partial p}{\partial z} = \rho_0 g\alpha G\zeta(\mathbf{x}), \tag{4.37a}$$

or, in dimensionless form

$$\frac{\partial P}{\partial Z} = \left\{ \frac{h^2}{l^2} \frac{k_H}{k_V} \right\} \frac{\zeta(\mathbf{x})}{h} \qquad \text{at } Z = 0. \tag{4.37b}$$

4.4.4 Two-dimensional reefs or banks

These geometries are the simplest. When the Y coordinate is taken along the bank, the middle term on the left of (4.36) is zero, and when $(l/h)^2 k_V/k_H \ll 1$ the coefficient of the third term is very small and can tentatively be neglected. The equation reduces simply to

$$\partial^2 P/\partial X^2 = -1,$$

The reduced pressure is a quadratic function of X, $P = -\frac{1}{2}X^2 + f(Z)$, where the arbitrary function $f(Z)$ depends on the shape of the bank. If the reef is symmetrical about $x = 0$, the solution with the physical units restored has the form

$$p = \frac{\rho_0 g\alpha G k_V}{2k_H}(l^2(z) - x^2), \tag{4.38}$$

where $l(z)$ is the half-width of the reef at a height z above the basement as in Figure 4.8 – $p = 0$ when $x = l(z)$.

This is a fine solution for the interior flow but, fairly obviously, it cannot be reconciled with the lower boundary condition (4.37a,b). The mathematical reason for the difficulty is our neglect of the vertical derivative term in (4.36), which reduces the order of the equation from second to zeroth in Z. As described in Section 4.3, we have a singular perturbation situation that points to the existence of an important lower boundary layer across which the *pressure* (in this instance) adjusts rapidly to connect the interior distribution (4.38) to the boundary pressure

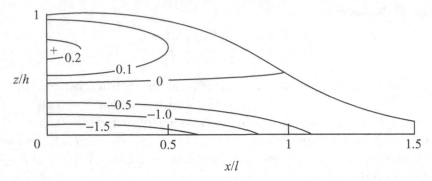

Figure 4.8. Contours of reduced pressure in units of $\rho g h^2 \alpha G$, for low Rayleigh number convection in a layered, two-dimensional bank, with the axis of symmetry on the left and $(k_H/k_V)(h/l)^2 = 25$. The corresponding streamlines in units of $\kappa(l/h)Ra$ are shown in Figure 4.9. Negative pressure near the base sucks fluid in from the flanks while the positive buoyancy above the base lifts the fluid against the pressure gradient, in an upwardly divergent flow.

gradient (4.37a,b). If the upper surface is flat, an upper boundary layer can also develop, as will be seen later.

The lower boundary layer is necessary not only mathematically (to complete the solution) but also physically, since it is necessary to drive the boundary layer flow towards the center. In the interior region above the boundary layer, the pressure is greatest along the axis of symmetry $x = 0$; its horizontal gradient drives the fluid *outward* with the velocity

$$u = -\frac{k_H}{\mu}\frac{\partial p}{\partial x} = \frac{g\alpha G k_V}{\nu}x, \qquad (4.39)$$

which increases linearly with distance from the symmetry axis. This is consistent with the more general conclusion following (4.33). Fluid is supplied to the interior by convective upwelling from below, from the lower boundary layer, but in order for the system to work, there must be a *negative* reduced pressure in the boundary layer near the bottom so that fluid can be sucked in from the sides before being lifted vertically by the buoyancy. The boundary layer pressure must then change rapidly across its thickness as the force balance near the basement (where the vertical velocity is small and the pressure gradient is established by the buoyancy) changes to that in the interior (where the buoyancy is balanced by the flow resistance, the vertical pressure gradient being smaller). The analysis of this lower boundary layer proceeds by the assignment to it of a local vertical scale δ, the boundary layer thickness in place of the bank thickness h, and insisting that in (4.36), the vertical

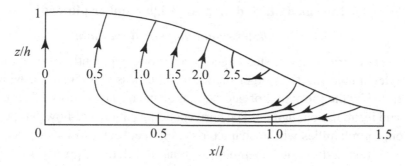

Figure 4.9. See the caption for Figure 4.8. Note the concentration of flow in the lower boundary layer.

pressure gradient term is balanced by the buoyancy force. This requires that

$$\delta \approx l \left(\frac{k_V}{k_H} \right)^{1/2}. \qquad (4.40)$$

The reason that the bank width determines the boundary layer thickness is that the interior flow is supplied by the lateral inflow in this region, as shown in Figure 4.9.

The solution can be completed analytically, but the details add little to the discussion above. The total volume flux Q per unit length of the bank can be estimated from the vertical flux out of the boundary layer, and it is found that

$$Q \approx \frac{g k_V \alpha G A}{\nu}, \qquad (4.41)$$

where A is the cross-sectional area of the bank. Note (i) that k_V is involved because the dominant resistance is from the vertical interior flow, not the faster, horizontal boundary-layer flow, and (ii) that the height of the bank is not involved, because the buoyancy and flow resistance are in local equilibrium in the interior.

In an axially symmetric bank with circular height contours and radius R, a similar boundary layer forms above the basement with radially inward flow and thickness $\delta \sim R (k_V / k_H)^{1/2}$ beneath the upward and outward interior flow. The total volume flux passing through the bank is

$$Q_T \sim \frac{g k_V \alpha G V}{\nu}, \qquad (4.42)$$

where V is the total volume of the bank. As in all of these low Rayleigh number flows, the velocity and volume flux are proportional to the Rayleigh number based on the geothermal flux with additional geometric factors.

4.5 Intermediate and high Rayleigh number plumes

4.5.1 Two-dimensional numerical solutions

Many thermally driven, quasi-steady geological flows apparently involve intermediate Rayleigh numbers (i.e. larger than 2 or 3 but less than 50, say), which are characterized by the formation of geothermal boundary layers, as well as internal plumes and recirculation regions. The corresponding patterns of flow, temperature, and so on, form families whose characteristics lie between those described in the previous section and various asymptotic forms of solution appropriate for high Rayleigh number flows. These are explored most conveniently by numerical analysis, although numerical calculations often require the choice of particular parameter values, and do not offer a clear sense of the ways in which the nature of the flow and reaction patterns change continuously with continuous changes in these parameters. In the second part of this section, simple scaling models are constructed from the internal balances in the flow to provide these algebraic connections. However, this kind of model can yield only an order of magnitude estimate of constants of proportionality, and to develop a useful quantitative insight it often requires the combination of the two approaches.

Consider a compact, two-dimensional, submerged, horizontally laminated platform of width l and height h with parallel escarpments on each side and a distributed heat flux from below. For two-dimensional Darcy flow in the x–z plane, the rotation vector expression (4.4) in the horizontal y-direction normal to the flow plane is

$$\Omega_Y = \frac{1}{k_V} \frac{\partial^2 \psi}{\partial x^2} + \frac{1}{k_H} \frac{\partial^2 \psi}{\partial z^2} = -\frac{g\alpha}{\nu} \frac{\partial T}{\partial x}. \tag{4.43}$$

This can be expressed in a form suitable for calculation by defining dimensionless horizontal and vertical coordinates X and Z, and temperature θ in terms of the natural scales of the system:

$$x = Xl, \quad z = Zh, \quad T(x, z) = T_0 \theta(X, Z), \tag{4.44a}$$

where T_0 is the maximum basement temperature relative to the ambient. The buoyancy force driving the flow acts vertically throughout the depth h of the platform, and the flow resistance necessarily involves the permeability k_V to vertical flow. Accordingly, the stream function ψ is expressed as

$$\psi(x, z) = \frac{g\alpha T_0 k_V h}{\nu} \varphi = \kappa(l/h) Ra \, \varphi(X, Z), \tag{4.44b}$$

where

$$Ra = \frac{g\alpha T_0 k_V h}{\nu\kappa} \tag{4.45}$$

is the Rayleigh number based on the vertical permeability and the platform depth. In terms of these variables, equation (4.43) becomes

$$\frac{\partial^2 \varphi}{\partial X^2} + \frac{k_V}{k_H}\left(\frac{l}{h}\right)^2 \frac{\partial^2 \varphi}{\partial Z^2} = -\frac{\partial \theta}{\partial X}, \tag{4.46}$$

where the aspect ratio l/h is usually large and the permeability ratio k_V/k_H usually small (< 1) in geological applications. From the definition of flow tube resistance (2.34), the *combination* of parameters $(k_V/k_H)(l/h)^2$ can be interpreted as the ratio of the vertical to horizontal flow conductances, which may be large (in a wide, un-layered platform, with $l \gg h$, $k_V \sim k_H$), or small in one that is narrow and heavily layered ($l \sim h$, $k_V \ll k_H$). The temperature field is specified by

$$u\frac{\partial T}{\partial x} + w\frac{\partial T}{\partial z} = \frac{\partial \psi}{\partial z}\frac{\partial T}{\partial x} - \frac{\partial \psi}{\partial x}\frac{\partial T}{\partial z} = \kappa\left\{\frac{\partial^2 T}{\partial x^2} + \frac{\partial^2 T}{\partial z^2}\right\}, \tag{4.47}$$

and in terms of the same dimensionless variables, this becomes

$$Ra\left\{\frac{\partial \varphi}{\partial Z}\frac{\partial \theta}{\partial X} - \frac{\partial \varphi}{\partial X}\frac{\partial \theta}{\partial Z}\right\} = \left(\frac{h}{l}\right)^2 \frac{\partial^2 \theta}{\partial X^2} + \frac{\partial^2 \theta}{\partial Z^2}. \tag{4.48}$$

Note that in (4.48), the horizontal and vertical advection terms on the left are comparable, but because the inverse aspect ratio h/l is usually very small ($\ll 1$), horizontal heat conduction is very small compared with that in the vertical direction. Note also that in this dimensionless form, the governing equations (4.46) and (4.48) and the boundary conditions below involve just three independent, dimensionless parameters, the scale ratio $(h/l)^2$, the permeability ratio (k_V/k_H) and the Rayleigh number Ra, which contain the nine physical quantities that can be counted. The nature of the solution therefore depends on the magnitude of these parameters only, rather than the nine physical quantities separately.

For the purpose of illustrating the evolution of the flow patterns as the parameters are varied, we will suppose that a submerged bank of rectangular cross section, surrounded by and saturated with saline water, rests on an impermeable basement at $Z = 0$. There is no flow across this lower boundary and the stream function $\varphi = 0$ there. The plane $X = 0$ is the axis of symmetry, so that $\varphi = 0$ also when $X = 0$. If the upper surface of the bank is exposed to the atmosphere, a shallow freshwater lens may form that inhibits flow across the surface, so that $\varphi = 0$ at $Z = 1$ also. If, however, the bank is awash or submerged, convecting fluid can move freely across its upper surface. The pressure at the upper surface is constant and fluid may rise vertically across it while the horizontal component of velocity along it is zero: $u = \partial \psi/\partial z = 0$ at $Z = 1$. In either case, fluid can enter or leave the bank horizontally at the side-walls at $X = \pm 2$, and the vertical velocity component $\partial \varphi/\partial X = 0$ there. In all of the calculations of this section, the scale l represents

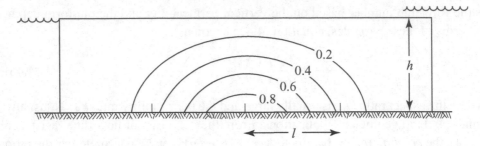

Figure 4.10. Purely conductive isotherms T/T_0 in a bank with a temperature distribution $T_0 \exp\{-(x/l)^2\}$ along the basement and zero along the other boundaries.

the half-width of the temperature distribution at the base of the convection region $T = T_0 \exp\{-(x/l)^2\}$, so that $\theta = \exp(-X^2)$ at $Z = 0$. The bounding temperature at the other interfaces is the ambient reference temperature, taken as zero and the distribution of isotherms when there is no flow is illustrated in Figure 4.10. The total width of the bank is $4l$, and since the flows are symmetrical about $X = 0$, only the right-hand half of each pattern is illustrated.

When the Rayleigh number is substantially less than unity, the flow field does not significantly disturb the temperature distribution of Figure 4.10, but as Ra increases, the isotherms become increasingly distorted by the flow – the temperature field and the flow pattern become more strongly interdependent. Figure 4.11 shows the results of calculations for Rayleigh numbers of 1 and 3, and Figure 4.12 for $Ra = 10$ and 30 when the upper interface is regarded as impermeable to the internal flow, being either capped by the water table or by the base of a freshwater lens, say. At each Rayleigh number, the three panels show streamline patterns, temperature distributions, and contours of the quantity $\mathbf{u} \cdot \nabla T = u(\partial T/\partial s)$. This quantity, called the rock alteration index (see Section 5.5), expresses the variation in the patterns and rates of reaction, precipitation or dissolution throughout the flow as the streamlines cut across the isotherms, and is described in more detail later. Calculations are shown at various values of Ra, taking $k_H/k_V = 10$ and $h/l = 1$. Compared with Figure 4.10 for $Ra \to 0$, the temperature field is distorted somewhat when $Ra = 1$, the isotherms being drawn inward near the base and upward in the upper region as a result of the flow. The rock alteration index is positive where fluid is moving to a higher temperature; if the equilibrium concentration increases with temperature, a positive alteration index indicates addition of solute to the interstitial fluid, and therefore, dissolution. A region of negative index indicates loss of solute from the interstitial fluid and therefore, deposition. For $Ra = 1$, the flow pattern is quite diffuse, driven upward about the axis of symmetry (the left boundary in the figure) by the distributed buoyancy, broadly outward near the top and inward above the

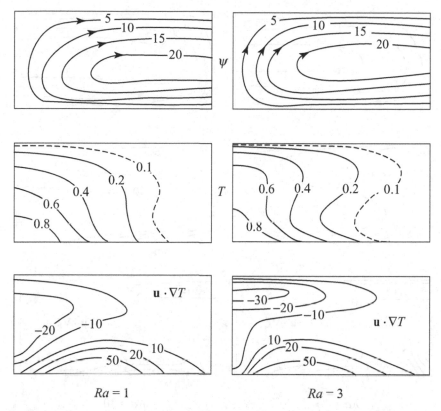

Figure 4.11. Convection patterns at Rayleigh numbers of 1 and 3, with no flow across the upper surface. Streamlines in units of $\kappa\,Ra(l/h) \times 10^{-1}$ are shown in the two top panels (the flow becomes more vigorous as Ra increases, since the units are larger in proportion), isotherms T/T_0 in the center panels and distributions of rock alteration index $\mathbf{u} \cdot \nabla T$, in units of $(\kappa T_0\,Ra/h^2) \times 10^{-1}$ in the lower ones.

bottom. When $Ra = 3$, this thermal distortion is greater; hot fluid near the axis is carried higher, so that the vertical temperature gradient along the axis is *reduced* in the interior and increased near the surface as a surface conductive layer develops that is not unlike that illustrated in Figures 4.8 and 4.9. On either side of the axis (only the right-hand side is shown), the vertical temperature gradient in the interior begins to reverse, as warmer fluid from the ascending stem turns outward, and cools. This is reflected in the patterns of the rock alteration index, $\mathbf{u} \cdot \nabla T$, where the more intense negative region (indicating deposition) migrates upward. As the Rayleigh number increases to 10 and 30 in Figure 4.12, the upward flow and the temperature distribution become more concentrated near the plane of symmetry, even though the basement heat flux and temperature distribution extend horizontally over the same distance as before. The basement is overlain by a thermal boundary layer,

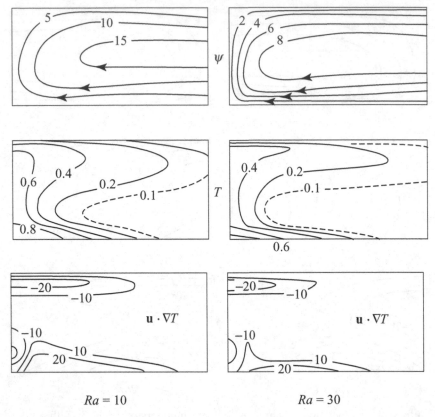

Figure 4.12. Convection patterns as in Figure 4.11, but for Rayleigh numbers of 10 and 30.

above which the temperature is close to that of the ambient water at the sides of the bank. The overall temperature pattern becomes increasingly more mushroom-shaped, with the lower thermal boundary layer merging into the central stem, then rising and spreading outward as an overhanging warm region, surrounding on three sides the nearly isothermal interior. Except near the bottom and the top, the isotherm shapes are close to those of the streamlines, so that the temperature gradient in the direction of flow becomes small, as is characteristic of high Rayleigh number flows in general (c.f. Section 4.3). Regions of significant negative values of the rock alteration index $\mathbf{u} \cdot \nabla T$ become confined to the lower part of the plane of symmetry where the fluid "turns the corner" and, more intensely, just below the upper surface in the upper thermal boundary layer where heat is conducted upward through the boundary and the vertical temperature gradient is large. At mid-depths in the moderately high Rayleigh number flows, the patterns of isotherms and streamlines are so similar that $\mathbf{u} \cdot \nabla T$ is small. Large positive and negative values of $\mathbf{u} \cdot \nabla T$ are confined to the thermal boundary layers above the basement and

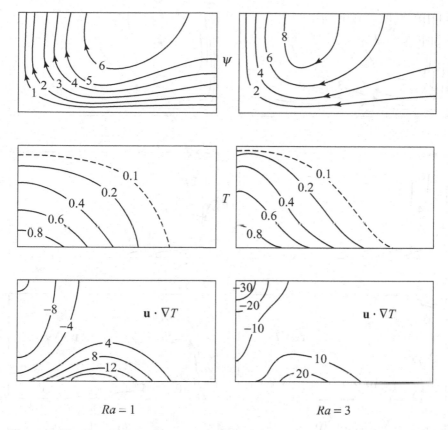

Figure 4.13. Convection patterns as in Figure 4.11 at Rayleigh numbers of 1 and 3 when the fluid is free to move across the upper and side surfaces. The scales are the same as in Figure 4.11.

below the upper surface, as fluid in these regions move essentially horizontally to higher and lower temperatures respectively.

Along the axis of the plume, the temperature decreases monotonically with height, but on the flanks if the flow is forced to spread laterally, the temperature along a vertical traverse downward has a maximum in the spreading region, below which it *decreases* until the lower thermal boundary layer is reached. The Florida Plateau seems to be the site of a natural circulation of this kind, as pointed out by Kohout (1965) and Kohout, Henry and Banks (1977).

Figures 4.13 and 4.14 show corresponding patterns found when the bank is totally submerged and the fluid is free to move across the upper surface. When $Ra = 1$, fluid is drawn into the bank mostly from the sides, but as Ra increases, the outward flow through the upper boundary and the temperature field gather increasingly into an upward plume above the temperature maximum. When the

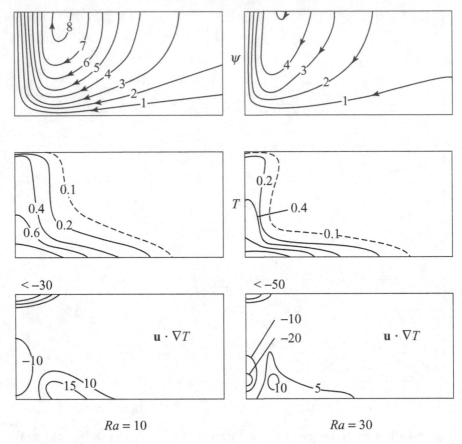

Figure 4.14. Convection patterns as in Figure 4.13 at Rayleigh numbers of 10 and 30, when fluid is free to move across the upper and side surfaces.

Rayleigh number is 3, the isotherms are again somewhat distorted, inward near the base and upward near the symmetry plane. When $Ra = 10$, the central truncated plume or stem structure begins to emerge, and by $Ra = 30$ it is clearly evident; since the fluid moves vertically across the upper surface, the plume is capped now by a convective–diffusive layer of the same genre as that in a pressure-driven discharge zone where the near-surface temperature gradient is large. The rock alteration index $\mathbf{u} \cdot \nabla T$ has very large negative values (indicating rapid deposition from solution) close to the surface over a horizontal region that corresponds to the plume width, and a region of moderately large positive values and dissolution just above the basement where fluid is being entrained into the hotter plume. In the interior, the intensity of gradient reactions is very much smaller, as is characteristic of high Rayleigh number flow. The horizontal extent of the regions of intense reaction near the upper surface is notably very much less than those just above the heated basement, and less than that found when the surface is impermeable to convective flow.

These calculations are intended to be illustrative but not exhaustive. The effect, for example, of variations in permeability ratio at large Rayleigh number can be inferred without further calculation. As k_H increases with all the other quantities remaining the same, the fluid moves horizontally much more readily than vertically, so that in the isothermal interior, the streamlines are more nearly horizontal and the lower thermal layer somewhat thinner. The central plume structure, including the distribution of rock alteration, is not altered much since there, the flow is essentially vertical and the overall distribution of $\mathbf{u} \cdot \nabla T$ is similar. In the interior region, the flow may be quite different, but the temperature gradients along streamlines are negligible, with much less rock alteration.

4.5.2 How do these flows work?

They are all driven by buoyancy forces associated with the temperature distributions, the connection being provided by the rotation vector relation (4.4), viz.

$$\Omega_Y = \frac{\partial u}{\partial z} - \frac{\partial w}{\partial x} = -\frac{k}{\nu}\frac{\partial b}{\partial x} = -\frac{kg\alpha}{\nu}\frac{\partial T}{\partial x}. \qquad (4.49)$$

The velocity scale for convective flows in general, (4.19), is determined by the ratio of buoyancy $g\alpha(\Delta T)$ to flow resistance ν/k. The temperature distributions can be represented accurately by the patterns of isotherms and two-dimensional flows by the patterns of streamlines. "Reading" these patterns in conjunction with each other, allows us to answer the question.

The simplest examples are shown in Figure 4.13 and 4.14, in which fluid can cross the upper bounding surface. At low Rayleigh numbers (1 and 3), the heat from the distributed basement source is transmitted largely by conduction through the matrix, and the isotherms are highest and horizontal at the axis of symmetry (the left-hand boundary in each panel) and all slope downward to the lower boundary, both on the right side (shown) and the left side mirror image. Near the axis of symmetry, the horizontal temperature gradient $\partial T/\partial x$ and the rotation Ω_Y are both very small, the flow is locally almost irrotational, and the nearly equally spaced streamlines show a uniform stream or jet rising vertically towards discharge across the upper boundary. Near the lower boundary, the (negative) horizontal temperature gradient is large and the velocity is negative (towards the left) and nearly horizontal ($w \ll u$) so that locally,

$$\frac{\partial u}{\partial z} \approx -\left(\frac{\alpha g k}{\nu}\right)\frac{\partial T}{\partial x} > 0.$$

The fluid speed towards the left is greatest at the lower boundary, then *decreases* with height above the basement, as confirmed by the increasing spacing with height of the streamlines in the lower-center of the top two streamline patterns. The broadly distributed temperature field at low Rayleigh numbers with streamlines intersecting isotherms, is associated with a broad distribution of the rock alteration index, although the characteristic values of the index itself, proportional to $\kappa T_0 Ra$, are relatively small. Significant degrees of matrix alteration require a correspondingly longer time.

At higher Rayleigh numbers, 10 and 30 in Figure 4.14, the streamline and isotherm patterns are qualitatively different. The more vigorous advection has swept the heat flux from the heated basement into a lower boundary layer and a central plume, with largely constant temperature, irrotational-flow regions filling the rest of the space. The flow patterns here have the same characteristics as the aquifer flows of Section 3.2, with gently curved, nearly uniformly spaced streamlines. The boundaries separating the rotational central plume and irrotational flow regions are determined by the balance between the convergent inflow carrying heat inward and the thermal diffusion outward; they lie somewhat below the rows of digits in the streamline patterns of Figure 4.14. At higher Rayleigh numbers, the inflow becomes relatively stronger and the diffusive regions (the central plume and the thermal boundary layer along the basement) both become thinner, as the isotherm patterns indicate. The flow in the central plume is very nearly vertical and uniform, with a direct local vertical balance between the buoyancy and the viscous retardation. Regions of significant matrix alteration are confined to the lower boundary layer and the lower and upper parts of the plume, where the streamlines intersect isotherms. In the central stem of the plume, the strong convection results in approximately parallel streamlines and isotherms and consequently small values of the rock alteration index.

Corresponding patterns of streamlines, isotherms and contours of the rock alteration index for the right-hand half of the platform are shown in Figures 4.11 and 4.12, for conditions in which the fluid cannot cross the upper surface. Again, the *horizontal* temperature gradient $\partial T / \partial x$ is negative and the rotation is clockwise throughout this side of the flow. Below the upper boundary, the isotherms are now deflected to the right by the discharge flow and at the larger Rayleigh numbers, the heated region develops an overhang there. The constant temperature, irrotational region in Figure 4.14 has disappeared, but again, the rock alteration index is small in the interior flow at the larger Rayleigh numbers because the streamlines and isotherms have similar shapes there. Most of the rock alteration occurs near the top and bottom of the platform as the streamlines pass through the thermal boundary layers.

4.5.3 Scaling analysis for two-dimemsional flows

A characteristic property of these solutions at even moderate values of the Rayleigh number is the existence of lower boundary regions where the temperature variations are large and flow is predominantly horizontal, converging toward the central plume. Their dominant characteristics can be found by scaling analysis. This is a technique for obtaining simple, approximate analytical expressions for the principal characteristics of complex fluid flows, such as (in this context) the parametric dependences of boundary layer thickness, plume width and total volume transport, etc., upon the geometry and the physical parameters contained in the dimensionless ratios $(h/l)^2$, (k_V/k_H) and Ra, above. With additional chemical parameters, such as the temperature dependence of solute saturation, $\partial c_E/\partial T$, rates of mineral redistribution can also be estimated.

Our interest here is to uncover parametric variations that are not evident from numerical calculations, which can give accurate results but only for a limited number of selected parametric values. The scaling analysis involves exploiting the flow geometry to replace the differential equations by algebraic approximations, which are much simpler to solve. Numerical coefficients in these balances (which are generally of order unity) are usually ignored, although in this particular example, it is worth while to recognize that the ratio k_H/k_V may be numerically large. Submerged permeable banks may be significantly stratified, with $k_H/k_V \gg 1$, so that the vertical and horizontal force balances are taken as

$$\mathbf{u} = (u, v) = -\frac{k_H}{\nu}\nabla_H p, \quad w = \frac{k_V}{\nu}\left(-\frac{\partial}{\partial z}(p/\rho_0) + b\right). \tag{4.50}$$

Consider cases in which Ra is at least moderately large and the bank is totally submerged, the interstitial fluid being free to vent through the upper surface. In the stem of the plume surrounding the central symmetry plane (the left-hand boundary in Figure 4.14), when $Ra > 10$, the streamlines and the isotherms are very nearly vertical. According to (4.44a,b), in the stem the buoyancy and the viscous flow resistance are very nearly in local balance, the vertical reduced pressure gradient being much smaller than either. In the plume center,

$$w_0 \sim \frac{k_V}{\nu}b_0 = \frac{k_V g\alpha T_0}{\nu} = \frac{\kappa}{h}Ra, \tag{A}$$

where this Rayleigh number involves the vertical permeability. In the corner near the symmetry plane and the basement, however, the vertical velocity is small and so is the viscous drag. Here, the fluid buoyancy $\rho g\alpha T_0$ acting upward must be balanced by a negative vertical gradient in reduced pressure. If the thermal boundary layer thickness there is δ, the central suction is

$$p \sim \rho g\alpha T_0 \delta,$$

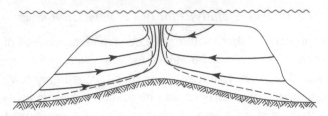

Figure 4.15. A schematic illustration of the flow in a submerged bank at high Rayleigh number, with a plume rising from a basement high. The heated region lies inside the broken lines, while outside, the temperature is close to that of the ambient seawater.

and from the Darcy balance, this produces an inward, approximately horizontal velocity equal to the product of $k_H/\mu = k_H \rho \nu$ and the mean horizontal pressure gradient (p/l). Thus close to the basement,

$$u \sim \frac{k_H g \alpha T_0 \delta}{\nu l} = \frac{\delta}{h} \frac{\kappa}{l} Ra_H, \quad \text{where} \quad Ra_H = \frac{g \alpha T_0 k_H h}{\kappa \nu}. \tag{B}$$

As is evident from Figure 4.14, the low pressure in this corner acts as a sink for the interior flow also, drawing in some additional cold fluid, which is heated as it is incorporated into the lower thermal boundary layer and the stem.

The volume flux of heated fluid in the thermal boundary layer, approximately $u\delta$, must be the same as that in (half of) the rising plume, i.e. $w_0 \, d$, where d is the plume half-width (see the cartoon Figure 4.15), so that

$$u\delta \sim w_0 d. \tag{C}$$

Finally, the basement heat flux is conducted across the lower thermal boundary at the rate (in kinematic units) $\kappa \, \partial T/\partial z \sim \kappa T_0/\delta$, where κ is the thermal diffusivity, and the total heat flux over the half-width l is approximately $\kappa T_0 l/\delta$. Since at large Ra, longitudinal heat conduction is negligible, this is equal to the heat convected by flow in the plume, that is, $w_0 d T_0$, so that $w_0 d T_0 \sim \kappa T_0 l/\delta$, and

$$w_0 d \sim \kappa l/\delta. \tag{D}$$

These four physical statements, (A)–(D), express the essential dynamics involved in this flow. We wish to determine the flow parameters u, d and δ in terms of the physical and geometrical properties of the region, the permeabilities and the Rayleigh number. The vertical velocity w_0 in the plume is already given by (A). From (D) and (C),

$$u \sim \kappa l/\delta^2,$$

which connects the inward velocity in the boundary layer to its thickness, δ. From substitution of this into (B), we obtain the lower boundary layer thickness scale as a fraction of the platform depth,

$$\frac{\delta}{h} \sim \left\{ \left(\frac{l}{h}\right)^2 \frac{k_V}{k_H} \right\}^{1/3} Ra^{-1/3} = (l/h)^{2/3} Ra_H^{-1/3} \tag{4.51}$$

in terms of a Rayleigh number containing the horizontal permeability, i.e. that for the direction in which the boundary layer fluid moves. At even moderately large Rayleigh numbers, then, the characteristic thickness of the lower boundary layer as a fraction of the height of the platform increases as the 2/3 power of the aspect ratio (l/h) and decreases as the inverse 1/3 power of the Rayleigh number. In Figure 4.14, this decrease in boundary layer thickness is becoming evident when Ra is 10 or 30, though the inverse one-third power variation could hardly have been extracted from the calculation. The result (4.51) is expected to fail when the aspect ratio (l/h) is very large, the structural stratification is weak and the upper surface is submerged. Under these conditions, the flow pattern is more compact, with fluid entering the platform across the upper surface over distances of the order of the layer depth h on either side of the central plume. In this limit, the flow pattern is independent of the platform width l and

$$\delta/h \sim Ra_H^{-1/3}. \tag{4.52}$$

From the thermal heat balance (D), above, the plume thickness

$$d \sim \kappa l/w_0\delta \sim \frac{l}{Ra_V} \left(\frac{l}{h}\right)^{-2/3} Ra_H^{1/3}, \qquad \text{from (A) and (4.50)}$$

and the width of the plume stem d, again referred to the platform depth, is

$$\frac{d}{h} \sim \left(\frac{l}{h}\right)^{1/3} \frac{k_H}{k_V} Ra_H^{-2/3}, \tag{4.53}$$

since

$$Ra_V = (k_V/k_H) Ra_H.$$

Note that in (4.53), field values of the ratio of the horizontal to vertical dimensions of the region and the permeability ratio are usually large so that even if the Rayleigh number is large, the stem of the two-dimensional plume circulation may be quite squat. In the calculations shown in Figure 4.14, the ratio $l/h = 1$ and $k_H/k_V = 10$. The decrease in plume width with increasing Rayleigh number, indicated by the crowding of the streamlines on the left (the center of the platform), is evident in the only when $Ra = 30$.

The geometrical patterns of mineral alteration in gradient reactions are seen from the patterns of $\mathbf{u} \cdot \nabla T$, but the total intensity of mineral alteration is determined by the vertical temperature gradient, the geochemistry of Section 5.5 and the total volume of solute that has passed through the system. From (D) above and equation (4.51) the volume of fluid F_V passing through the system per unit time per unit length of the platform containing the two-dimensional plume is

$$F_V \sim wd \sim \kappa l/\delta \sim \kappa \left\{ \frac{l}{h} Ra_V \right\}^{1/3} \tag{4.54}$$

for moderately large values of the Rayleigh number, and the total volume of fluid per unit area that has passed through is given by this expression multiplied by the time duration of the flow. The numerical coefficient of proportionality can be found by fitting (4.54) to the results of numerical calculation and it is found to be about 0.3.

4.5.4 Circular platforms

Wilson *et al.* (1990) have given a similar scaling analysis for axially symmetrical, high Rayleigh number flows, considering a circular permeable platform of radius R_P and height h, heated from below, as before (see Figure 4.16 later). As in the previous example, a few basic physical balances determine the dominant characteristics of the flow. First, the almost vertical flow in the stem is again in close local balance between the buoyancy and the viscous flow resistance, so that the vertical velocity along the center of the stem,

$$w_0 \approx \frac{k_V g \alpha T_0}{\nu} = \frac{\kappa}{h} Ra_V. \tag{4.55}$$

Next consider the heat balance, which equates the heat convection upwards in the plume with a stem of radius R_S, i.e. $w_0 T_0 \cdot \pi R_S^2$, to the heat conducted upward from the basement into the lower boundary layer, $(\kappa T_0/\delta)\pi R_P^2$. Thus

$$w_0 \approx \frac{\kappa}{\delta} \left(\frac{R_P}{R_S} \right)^2. \tag{4.56}$$

The total volume flux of fluid Q up the stem is $\pi R_S^2 w_0$, so that

$$Q \sim \frac{\kappa}{\delta} R_P^2. \tag{4.57}$$

Also, elimination of w_0 between (4.55) and (4.56) defines the Rayleigh number of the flow in terms of flow dimensions that are potentially measurable in the field

$$Ra_V \sim \left(\frac{R_P}{R_S} \right)^2 \frac{h}{\delta}, \tag{4.58}$$

where δ is the thickness of the thermal boundary layer near the stem whose radius is R_S.

These simple results are very useful for geological interpretation of paleo-convection sites such as the Latemar Massif in the Italian Dolomites, discussed in some detail in Section 5.8. The thermal diffusivity κ for most rocks is in the range $(4\text{--}10) \times 10^{-3}$ cm^2/s or $(4\text{--}10) \times 10^{-7}$ m^2/s. If the platform radius can be estimated from the regional geology and the lower thermal boundary layer thickness and stem radius from the patterns of geochemical alteration, the vertical velocity in the stem is given by (4.54), and the Rayleigh number characterizing the flow when it was active, is found from (4.56) without further speculation.

4.5.5 Similarity solutions – two-dimensional plumes

Consider a saturated, semi-infinite, permeable region above an impermeable basement with a long, narrow heat source. The thermally induced convective flow can be considered two-dimensional in the transverse plane, and the stream-function representation $u = \partial\psi/\partial z$, $w = -\partial\psi/\partial x$ is used to specify the flow. Thus,

$$\frac{\partial\psi}{\partial x} = -\frac{gk\alpha T}{\nu} \tag{4.59}$$

and

$$\frac{\partial\psi}{\partial z}\frac{\partial T}{\partial x} - \frac{\partial\psi}{\partial x}\frac{\partial T}{\partial z} = \kappa\frac{\partial^2 T}{\partial x^2}. \tag{4.60}$$

An important, but unsurprising, result that can be established immediately is that the convective heat flux in the plume, namely,

$$F_{\mathrm{H}} = (\rho C)_{\mathrm{F}} \int_{-\infty}^{\infty} wT\,dx \tag{4.61}$$

is constant with height and equal to the total rate of heat input at the bottom. This may be intuitively obvious, but can be shown formally shown by integration of (4.60), which can be written as

$$\frac{\partial}{\partial x}\left(\frac{\partial\psi}{\partial z}T\right) - \frac{\partial}{\partial z}\left(\frac{\partial\psi}{\partial x}T\right) = \kappa\frac{\partial^2 T}{\partial x^2}, \tag{4.62}$$

so that

$$-\frac{\partial}{\partial z}\int_{-\infty}^{\infty}\frac{\partial\psi}{\partial x}T\,dx = \int_{-\infty}^{\infty}\frac{\partial}{\partial x}\left(\kappa\frac{\partial T}{\partial x} - \frac{\partial\psi}{\partial z}T\right)dx = 0, \tag{4.63}$$

since $T \to 0$ as x $\to \pm\infty$ at all z. Consequently,

$$-\int_{-\infty}^{\infty} \frac{\partial \psi}{\partial x} T \, dx = \int_{-\infty}^{\infty} wT \, dx = \frac{F_H}{(\rho C)_F} = Q, \text{ say}; \tag{4.64}$$

the total heat flux Q is independent of z, as anticipated above. The vertical velocity is proportional to the temperature relative to the ambient so that this last equation provides an integral constraint on the temperature distribution:

$$\int_{-\infty}^{\infty} T^2 dx = \frac{\nu Q}{k_V g \alpha}, \tag{4.65}$$

which, again, is independent of height.

Wooding (1963) demonstrated that equations (4.61) and (4.62), subject to the condition (4.65), allow similarity solutions, that is, solutions that are functions of a single dimensionless combination of the physical variables, so that the partial differential equations become ordinary differential equations and the solution profiles have the same shape at each level. The procedure for finding similarity expressions is explained in books on fluid mechanics, such as that of Batchelor (1967). In this context, they are of the form

$$\psi = -\kappa(z/\lambda)^{1/3} f(\eta), T = (Q/\kappa)(\lambda/z)^{1/3} f'(\eta), \tag{4.66}$$

where the parameter

$$\lambda = \frac{\nu \kappa^2}{g \alpha k_V Q} \tag{4.67}$$

is an intrinsic length scale for the plume (i.e. one determined by the physical parameters of the flow, not the geometry) and the similarity variable in (4.66) is

$$\eta = \frac{x}{(\lambda z^2)^{1/3}}. \tag{4.68}$$

Since $T \to 0$ as $x \to \pm\infty$ (far outside the plume),

$$f'(\eta) \to 0 \quad \text{as} \quad \eta \to \pm\infty. \tag{4.69}$$

Clearly $f(\eta)$ is constant outside the plume (though it has different values on the two sides because of the upward net flux in the plume), so let

$$f(\eta) \to 0 \quad \text{as} \quad \eta \to \infty, \quad f(\eta) \to -c \quad \text{as} \quad \eta \to -\infty, \tag{4.70}$$

where the constant c is to be determined. Finally, the condition (4.65) requires that

$$\int_{-\infty}^{\infty} f'^2(\eta)d\eta = 1. \tag{4.71}$$

The form of the function $f(\eta)$ is found by substitution of (4.66) into (4.62), whence after reduction,

$$ff'' + f'^2 + 3f''' = 0. \tag{4.72}$$

This nonlinear equation occurs in the analysis of a viscous two-dimensional jet in fluid mechanics (see, for example, Batchelor, 1967, p. 345) and can be integrated exactly. The solution is

$$f(\eta) = c\tanh(c\eta/6). \tag{4.73}$$

The constant c is determined by the condition (4.71); after a small calculation it is found that $c = (9/2) \approx 1.65$. Accordingly, the solutions for the stream function and the temperature distribution are

$$\begin{aligned}\psi &= -c\kappa(z/\lambda)^{1/3}\tanh(c\eta/6),\\ T &= \frac{c^2}{6}\left(\frac{Q}{\kappa}\right)(\lambda/z)^{1/3}\text{sech}^2(c\eta/6).\end{aligned} \tag{4.74}$$

These are illustrated in Figure 4.16.

The total volume flux per unit horizontal length of the plume is the difference between the values of ψ at $\eta = +\infty$, that is, $3.3\kappa(z/\lambda)^{1/3}$. Note that the internal length scale λ depends not on any geometrical length, but on the physical parameters of the situation as in (4.67). The plume width beyond which the temperature and velocity fields drop off rapidly, is $\delta \approx 7.3(\lambda z^2)^{1/3}$, increasing with height (as $z^{2/3}$). The temperature and the vertical velocity along the centerline decrease rather slowly (as $z^{-1/3}$), while the total volume flux increases as $z^{1/3}$ as a result of fluid being drawn into the plume from either side. It will be shown in the next chapter that the quantity $u \cdot \nabla T$, the rate at which fluid is moving up the temperature gradient and called the rock alteration index, is a measure of the reaction rate in flow controlled, gradient reactions. In a plume of this kind, the general magnitude of the rock alteration index is $\kappa T/\delta^2$, where T is the axial temperature and δ the plume half-width; it is positive on the flanks of the plume as cold water is drawn into it and negative on the plane of symmetry as fluid moves upward, down the temperature gradient.

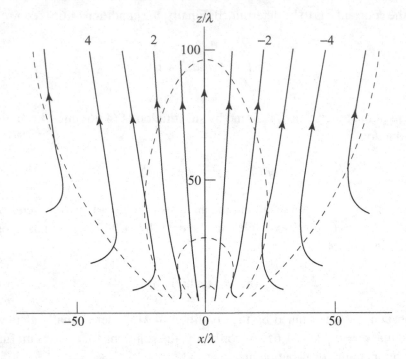

Figure 4.16. A two-dimensional, high Rayleigh number plume in a porous medium above a line source of heat along $x = z = 0$, from Wooding (1963). The streamlines are in units of κ, and the isotherms, shown as dashed lines, are at intervals $0.05\, Q/\kappa$. Note how squat this pattern is, compared with a typical atmospheric or laboratory plume.

4.5.6 The axi-symmetrical plume in a semi-infinite region

When the plume rises from a localized maximum in the basement heat flux or a local basement high, it may be approximately axi-symmetrical, and solutions were given by Wooding (1963) for this case also. The flow can now be represented in terms of the Stokes stream function (2.16),

$$u = \frac{1}{r}\frac{\partial \psi_{\mathrm{S}}}{\partial z}, \quad w = -\frac{1}{r}\frac{\partial \psi_{\mathrm{S}}}{\partial r}, \tag{4.75}$$

where r and z are the radial and vertical coordinates, respectively. The vertical velocity is again proportional to the temperature relative to the ambient, as in (4.43), and the cross-sectional profiles of temperature and vertical velocity are similar

$$w = -\frac{1}{r}\frac{\partial \psi_{\mathrm{S}}}{\partial r} = \frac{g k_{\mathrm{V}} \alpha T}{\nu}. \tag{4.76}$$

The thermal energy balance becomes

$$\frac{\partial \psi_S}{\partial z}\frac{\partial T}{\partial r} - \frac{\partial \psi_S}{\partial r}\frac{\partial T}{\partial z} = \kappa \frac{\partial}{\partial r}\left(r\frac{\partial T}{\partial r}\right). \tag{4.77}$$

The constancy of the vertical convective heat flux in the plume can be established as before by integration over a horizontal plane intersecting the plume:

$$2\pi \int_0^\infty wTr\,dr = Q = \text{const}, \tag{4.78}$$

where $(\rho C)_F Q$ is the total heat flux in the plume (not the heat flux per unit length as it was in the two-dimensional case previously).

Similarity solutions can again be found for this case. They are of the form

$$\psi_S = \kappa z f(\eta), \quad T = \frac{Q}{\kappa z}g(\eta), \tag{4.79}$$

where the similarity variable, the scaled radius, is

$$\eta = \left(\frac{gk_V\alpha Q}{\nu\kappa^2}\right)\frac{r}{z} = \lambda_A^{-1}(r/z), \tag{4.80}$$

where, as suggested by the notation, λ_A is closely related to the length scale λ of (4.67), but is here based on the total heat flux in the plume. The precise forms of the profiles f and g are found by substitution of (4.80) into (4.76) and (4.77) (Wooding, 1963) and are illustrated in Figure 4.17.

The properties of the axi-symmetrical solutions differ somewhat from those for the two-dimensional case. The half-width of the plume increases linearly with height z above the origin

$$\delta \approx 16(Ra)^{-1/2}z,$$

so that the region occupied by the plume resembles an inverted cone whose semi-angle decreases as the Rayleigh number increases. The vertical velocity along the axis decreases inversely with height in spite of a radial inflow proportional to κ/r; the axial heat flux is, of course, constant. The temperature along the vertical axis decreases inversely with height, more rapidly than the two-dimensional case because the heat is advected and diffused in all horizontal directions, rather than just on both sides.

Natural geothermal flows often occur along a segmented rift system such as the Taupo Volcanic Zone in the North Island of New Zealand, discussed extensively by Rowland and Sibson (2004). This is a region of active extension and productive volcanism, having a vertical heat flux through the central part of the zone that is

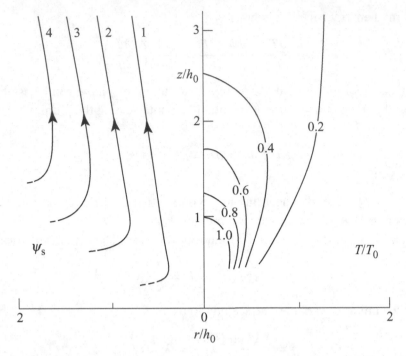

Figure 4.17. The axisymmetric plume above a point heat source at $Ra = 100$. The temperature scale T_0 is the temperature on the axis at an arbitrarily chosen height h_0. The distribution of Stokes stream function ψ_S on the left has units κh_0; equal increments of ψ_S specify equal increments of total transport through the axi-symmetrical volume contained.

ten times the average continental heat flux. The geothermal fields have convective circulations, possibly similar to Wooding's two-dimensional plume structures described above, but extending to depths of 7–89 km. Parallel normal faults, fractures and dikes dissect the convective flow regime and either enhance or restrict the flow according to the relative permeability of the structure and the host rock. The extensional fabric is partitioned into discrete rift segments, connected by accommodation zones. Rowland and Sibson point out that the maintenance of structural permeability requires repeated brittle failure for the geothermal plumes to continue, and that the pattern of flow is determined by the architecture of the flow paths, not by the hydrodynamics.

4.6 Salinity-driven flows

There are two important dynamical differences between flows where the buoyancy is the result of temperature differences and those produced by salinity variations. Since saline water is essentially confined to the interstices, the advection velocity for salt or other dissolved components is the mean interstitial velocity, whereas heat diffuses into the solid matrix and the advection velocity for heat is approximately the same as the transport velocity. Secondly, the thermal diffusivity κ in the heat balance is a material property of the matrix, while the dispersion of salt in solution results predominantly from the structural randomness that exists in natural permeable materials. The solute dispersion coefficient D is proportional to the interstitial flow speed and is of the form $\bar{v}\alpha_D$, where α_D is the dispersivity defined in Section 2.10 and $\bar{v}$ is the mean interstitial speed. Since both transport processes, solute advection and solute dispersal, are proportional to the interstitial fluid speed, the balance between them depends only on the flow geometry, not on the flow magnitude.

4.6.1 Freshwater lenses

A freshwater lens above a saline substrate offers an example in which the distribution of salinity S is an important determinant of the flow. Horizontal temperature differences are likely to be small. In aquifers adjacent to saline water bodies such as Long Island, New York, or many coral-based islands in deep water, rainwater infiltration into the aquifer water table displaces denser interstitial saltwater below the land surface, forming a freshwater lens. The upper surface of the lens is highest along the groundwater divide and the internal flow discharges from the circumference of the lens along the shore lines. Beneath the lens, the permeable substrate is saturated with denser saline water usually over an impermeable basement, and the outward freshwater flow is driven by the downward slope of the water table towards discharge. Correspondingly, the lower surface of the lens is depressed by the weight of the freshwater between water table and the seawater level outside. Because of the approximate overall hydrostatic balance, the shape of the fresh–salt water interface generally reflects that of the water table above, though inverted and magnified, and possibly modified by the existence of low permeability, internal retarding layers. The density contrast across the lower fresh–saline water interface is characteristically about 3%, so that this interface extends about 30 times deeper below sea level than the water table extends above it. In spite of this magnification, the slope of the fresh–saline water interface usually remains numerically small.

This interface, or transition layer, separating the more buoyant, fresh interstitial water from the saline-saturated region below is an essential part of the circulation pattern. It is deepest beneath the highest points of the freshwater table near the center of the lens and slopes upward toward the discharge along its circumference, as shown in Figure 4.18. From the isothermal equation of state (4.2) with $T = 0$, the buoyancy of the fresh water in the lens relative to the salt water below is $g\beta S$. The upward slope θ of the layer produces the *horizontal* buoyancy gradient generating

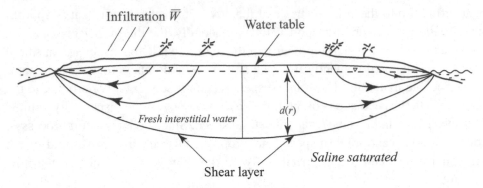

Figure 4.18. A cartoon illustrating a freshwater lens beneath a permeable island with surface infiltration, with the vertical scale exaggerated. The configuration of the water table is reflected and amplified in the lower boundary shear layer, separating regions of outwardly moving freshwater above and relatively stagnant saltwater below. The slope of the shear layer and the density difference above it balances the horizontal pressure gradient above, so that the interstitial seawater in the medium below is in essence stagnant.

rotation and shear between the irrotational outward flow above and the essentially stagnant seawater below, specified in two-dimensional flow by (4.4),

$$\Omega_Y \approx \frac{\partial u}{\partial z} = -\frac{k}{\nu}\frac{\partial b}{\partial x} = K\beta\frac{\partial S}{\partial z}\sin\theta, \qquad (4.81)$$

where $K = gk/\nu$ is the hydraulic conductivity. Note the correspondence between the profiles of up-slope velocity and buoyancy in the transition layer. On integration, we have

$$u(z) \approx K\beta\{S_0 - S(z)\}\sin\theta, \qquad (4.82)$$

where S_0 is the salinity in the essentially stagnant interstitial fluid below the layer. Just above the transition layer, in the aquifer flow, $S(z) = 0$ and

$$u \approx K\beta S_0 \sin\theta. \qquad (4.83)$$

The same relation holds in the axi-symmetrical case, with u now being the radial velocity.

The profiles of the water table above sea level (and of the fresh–saline water interface below) can be found simply using the small slope approximation of the previous chapter. The effective hydraulic boundary of the flow is the line of intersection of the water table and the fresh–saline interface; in nature, it is necessarily offshore since freshwater may vent from the permeable medium both above sea level from the banks and below, from the near-shore sea-bed. In the case of a circular island of effective radius R enjoying uniform infiltration $\overline{W}$, conservation of water mass requires that, at any radius $r < R$, the radial outflow of freshwater balances the total infiltration from above,

$$u \cdot 2\pi r d(r) = \pi r^2 \overline{W},$$

where $d(r)$ is the depth below sea level of the fresh–salt water transition layer at radius r. To the small slope approximation, (4.83) can be expressed as

$$u \approx -K\beta S_0 \frac{\partial d}{\partial r}.$$

The elimination of u between these two equations gives the lower profile of the freshwater lens,

$$d(r) \approx \left\{ \frac{\overline{W}}{K\beta S_0} \right\}^{1/2} \left(R^2 - r^2\right)^{1/2}. \tag{4.84}$$

The height of the water table above sea level has the same form but is smaller in magnitude by a factor of $\delta\rho/\rho = \beta S_0 \sim 0.03$. The maximum depth of the freshwater lens below sea level is at $r = 0$:

$$d(0) = \left\{ \frac{\overline{W}}{K\beta S_0} \right\}^{1/2} R. \tag{4.85}$$

Tidal oscillations along a steep shore line may interact with the water table configuration to a distance $K/\phi n$, where n is the diurnal tidal frequency, $2\pi/$ (1 day), generally of order 100 m, or less. A more interesting tidal rectification process occurs in shallow embayments with extensive beaches and sandbars that are exposed at low tide. The saline tidal inflow at high tide "tops up" the water table in the sandbanks along the shoreline and sinks into the sandbar on the ebbing tide. The relaxation time (Section 3.2) for drainage from the interior of the sandbar is

$$T_{\text{RX}} = \frac{\phi R^2}{2K d(0)} \sim \frac{\phi d(0)}{\overline{W}},$$

which is generally measured in weeks and is always much larger than the tidal period, so that the water table has relaxed only slightly from the previous high tide level when the next arrives. As a consequence, the mean height of the water table beyond the beach is significantly higher than the mean sea level and almost as high as the highest diurnal or semi-diurnal tides. This tidal rectification effect has been observed in a sandy spit south of the outflow of the Hawkesbury River in New South Wales. Continuing freshwater outflow leads to the locally unstable situation described later in this chapter, with saline water of the tidal inflow over the fresh interstitial water in the groundwater outflow and consequent local vertical mixing.

Inland of the shoreline and above the tidal range, however, the groundwater near the surface is fresh and the density difference across the transition layer is dynamically stable with no obvious interior energy source to produce mixing along the interface that defines the base of the freshwater lens. It is expected to remain relatively sharp unless the medium is fractured or traversed by extensive conduits. In any event, the mean flow in the transition layer is outward, so that mixed fluid arriving near the shoreline moves generally towards the sea.

4.6.2 Gravity currents in porous media

The term "gravity current" is used to describe a flow that occurs when a denser fluid intrudes into a fluid domain and, because of its greater density, spreads primarily horizontally over the lower boundary of the domain.

Some of the early research on buoyancy driven flow in porous media has been summarized in Bear's (1979) classic text. More recently. Huppert and Woods (1995), Woods and Mason (2000) and Lyle *et al.* (2005) have conducted interesting laboratory experiments that are related to the flow in a freshwater lens described above. In a natural freshwater lens, lighter water enters uniformly across the upper boundary, spreads radially and discharges at the fixed circumference, while in Lyle *et al.*'s experiment, the geometry is inverted. Denser fluid starts to flow from an orifice at the base of a laboratory tank filled with a porous medium saturated with less dense water. The denser saline water forms an axi-symmetric gravity current that deepens and spreads across the floor of the tank. In each experimental run, the volume flux of fluid, Q, commences at some initial time and thereafter remains constant; its density is $\rho + \Delta\rho$ and that of the surrounding interstitial fluid is ρ. Initially, the intruding fluid was observed to flow from the orifice in all directions because the internal Froude number $u^2/(\Delta\rho/\rho_0)gR$ is initially large, where u is the outflow speed, $(\Delta\rho/\rho_0)g$ is the fractional buoyancy difference and R is the flow radius. However, as the radius increases, the density difference causes the intrusion to spread horizontally more rapidly than it does vertically. The height of the intruding body of fluid as a function of radius r in cylindrical coordinates is represented as $h(r, t)$ and one series of measured profiles is shown in Figure 4.19.

When the interfacial slope has become small, the internal pressure distribution is approximately hydrostatic in the vertical. The radial pressure gradient

$$\frac{\partial p}{\partial r} = g(\Delta\rho)\frac{\partial h}{\partial r}$$

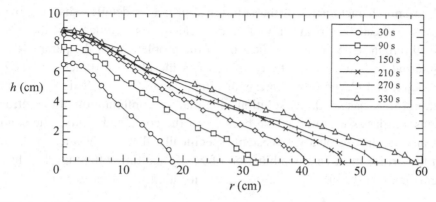

Figure 4.19. The measured height profiles of an axi-symmetrical gravity current as a function of radius and time, but with $g' = (\delta\rho/\rho)g = 40\,\text{cm/s}^2$ and volume flux $Q = 34.4\,\text{cm}^3/\text{s}$, as measured by Lyle *et al.* (2005).

is independent of height in the intrusion and this, in turn, generates a horizontal radial velocity

$$u(r) = -\frac{k}{\mu}\frac{\partial p}{\partial r} = -K\frac{\Delta\rho}{\rho}\frac{\partial h}{\partial r}, \qquad (4.86)$$

where K is the hydraulic conductivity of the medium. This velocity is a function of radius and time, but is independent of height in the intruding fluid above the floor of the tank. The incompressibility condition (2.15) can be integrated from the tank floor to the interface at $z = h$, and this gives the vertical velocity of the interface:

$$w(h) = \phi\frac{\partial h}{\partial t} = \frac{1}{r}\frac{\partial}{\partial r}(ruh), \qquad (4.87)$$

where ϕ, assumed constant, is the porosity. Substituting (4.86) into (4.87), we obtain the nonlinear partial differential equation governing the interfacial height $h(t)$ of the intruding region:

$$\frac{\partial h}{\partial t} - \frac{K'}{r}\frac{\partial}{\partial r}\left(rh\frac{\partial h}{\partial r}\right) = 0, \qquad (4.88)$$

where

$$K' = \phi^{-1}(\Delta\rho/\rho)K \qquad (4.89)$$

is a form of the hydraulic conductivity for internal, porous media flow. The solution for $h(r, t)$ is subject to the condition that the volume of the intruding fluid is equal to the total efflux from the orifice:

$$2\pi\int_0^{r_N(t)} rh\,dr = Qt, \qquad (4.90)$$

where $r_N(t)$ is the radial position of the nose of the intrusion, where $h = 0$.

The main characteristics of the solution (but not its detailed shape, shown in Figure 4.19) can be found without further calculation. Notice that there is no physical length scale in the specification of the problem, only the parameters K', the internal hydraulic conductivity, which has the physical dimensions $[LT^{-1}]$ of velocity, and Q, the influx rate with dimensions $[L^3 T^{-1}]$. This suggests that equation (4.88) has similarity solutions, i.e. that the configuration of the intrusion remains similar as it grows in size because of the continued influx. The solution must also be dimensionally consistent, specifically, that the distance of the nose from the orifice must be a combination of K', Q and time t that has the physical dimensions of a length. Consequently, the location of the nose must be of the form

$$r_N(t) \propto (K'Q)^{1/4} t^{1/2}, \qquad (4.91)$$

spreading as $t^{1/2}$. The area over which the intrusion has spread is then proportional to t, i.e. to the total volume injected, so that when the slope of the interface has become small, the mean vertical extent of the intrusion remains the same during the subsequent spreading.

The laboratory measurements by Lyle and her colleagues (2005) confirmed these expectations. The measured shape of the gravity current at different times (Figure 4.19) shows the central height increasing rapidly at first, and then stabilizing as the interfacial slope reduces and the flow moves toward the similarity form (4.91). The advance of the nose was found to follow (4.91) very closely over wide time intervals with a constant of proportionality of 0.51.

4.7 Thermal instabilities

Instability can arise in porous media flows when variations in fluid density are present as a result of variations in temperature, salinity, or both. A clear distinction must be drawn between, on the one hand, buoyancy-driven flows that *necessarily* occur when the interstitial fluid isopycnals slope relative to the horizontal and, on the other hand, the self-generating and self-patterning instability flows that may develop from a state of rest or steady motion even when the isopycnals *are* horizontal. For example, a local hot spot at the basement of a horizontal permeable layer generates a convective circulation above it whose strength is roughly proportional to the temperature excess. In contrast, above a uniformly heated horizontal basement, interstitial fluid can remain in equilibrium at rest, with the heat flux being purely conductive. If, however, the Rayleigh number associated with the *uniform* heating becomes too large, an organized cellular motion can appear and grow as a result of an internal instability, so that the self-generated

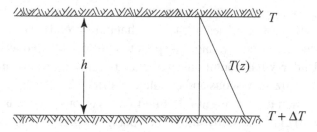

Figure 4.20. The initial steady state at rest with a Newtonian fluid between the plates, as considered by Rayleigh (1916), or with a saturated porous medium, considered by Lapwood (1948).

convective heat flux becomes significant. It is important to be able to recognize the conditions under which these instabilities can occur.

The geometries that we consider are usually simplifications of nature: interfaces between different regions are never quite smooth, nor the regions themselves quite homogeneous – random perturbations abound. The general question of dynamical stability concerns the response of the flow patterns and temperature distributions to such perturbations: are the temperature and flow perturbations suppressed or smoothed out, or do they amplify and grow to the extent that they seriously modify the original state, which then moves to a different, more stable configuration?

Mathematical solutions that, when disturbed by a certain class of perturbations, always return to the original solution are said to be dynamically stable to those perturbations. If, however, *any* member of the class of perturbations is found to amplify in time (even though all others may decay), the solution is unstable.

The technique for deciding whether a given solution is stable (or under what conditions it is stable), involves three essential steps. (i) The basic state that we wish to examine, the distribution of flow, temperature, etc., must be specified as a solution to the governing equations and the appropriate boundary conditions. (ii) The solution is perturbed slightly in as general a manner as possible consistent with the boundary conditions. The combined basic state and perturbations must also satisfy the governing equations. Finally, (iii) we follow the time evolution of the perturbations to determine the conditions under which they all decay or fail to grow, that is, the conditions for stability.

4.7.1 Rayleigh–Darcy instability

The archetypical thermal stability situation is illustrated in Figure 4.20. Consider a horizontal layer of viscous, heat-conducting fluid confined between parallel plane surfaces with separation h. The temperature of the lower surface is higher by the amount ΔT than the upper surface; in the steady state at rest there is a uniform heat flux $\kappa \Delta T / h$ upward. Now, the fluid near the lower surface is hotter, less dense, and so more buoyant than the fluid near the top, so that it may tend to rise. However, fluid motion is resisted by viscosity, and hot spots tend to be diffused away by thermal conductivity. Under what conditions will the destabilizing buoyancy variations overcome the stabilizing viscous and conductive effects?

The case in which the gap is entirely filled with viscous fluid was the subject of Rayleigh's (1916) classical study; the somewhat more pertinent situation in which the gap is occupied by a porous medium saturated with an interstitial fluid of uniform composition was solved by Horton and Rogers (1945) and in more detail by Lapwood (1948). The equations governing any motion in the layer are Darcy's equation, the incompressibility condition, the heat balance without internal heat sources, and the equation of state for a homogeneous fluid:

$$\nabla_H p = -\frac{\mu}{k} \mathbf{u}_H,$$

$$\frac{\partial p}{\partial z} = -\frac{\mu}{k} w - \rho g,$$

$$\nabla \cdot \mathbf{u} = 0, \tag{4.92}$$

$$M \frac{\partial T}{\partial t} + \mathbf{u} \cdot \nabla T = \kappa \nabla^2 T,$$

$$\rho = \rho_0 (1 - \alpha T),$$

where $M = (\rho C)_M / (\rho C)_F \sim 0.5$ is the matrix-to-fluid specific heat ratio in (2.48), κ is the thermal diffusivity and $\mathbf{u} = (\mathbf{u}_H, w)$. The mass balance for dissolved material is satisfied trivially since $c = $ const. At the lower boundary $z = 0$, the normal velocity component $\mathbf{u} \cdot \mathbf{n} = w = 0$ and the temperature $T = T_0 + \Delta T$, while at the top boundary, $z = h$, we have that $w = 0$ and $T = T_0$. The steady state of rest with uniform temperature gradient is

$$\mathbf{u}_H = w = 0, \quad p = p_0(z), \quad \partial p_0 / \partial z = -\rho g, \tag{4.93}$$

$$T = T_0 + \left(1 - \frac{z}{h}\right) \Delta T.$$

The expressions (4.93) satisfy the equations above without restrictions on the parameters, but under what condition is this state of rest stable?

To examine this question, we do what an experimentalist would do – disturb the basic state slightly. The small perturbations are marked with primes.

$$\mathbf{u} = (\mathbf{u}_H, w) = \mathbf{u}'(x, y, z, t),$$

$$p = p_0(z) + p'(x, y, z, t), \tag{4.94}$$

$$T = T_0 + \left(1 - \frac{z}{h}\right) \Delta T + T'(x, y, z, t).$$

The now evolving flow, temperature, and pressure distributions must also satisfy (4.92). Substitution of (4.94) into (4.92) gives, after subtracting out the basic state pressure balance,

$$\frac{\partial p'}{\partial z} = -\frac{\mu}{k}w' + \rho_0 g\alpha T',$$
$$\nabla_H p' = -\frac{\mu}{k}\mathbf{u}'_H, \qquad\qquad (4.95)$$
$$\nabla \cdot \mathbf{u}' = 0,$$

and

$$M\frac{\partial T'}{\partial t} + \mathbf{u}' \cdot \nabla T' - w'\frac{\Delta T}{h} = \kappa\nabla^2 T',$$

with $w' = T' = 0$ on $z = 0$ and h. The equation of state has here been incorporated into the last term of the first equation.

These equations govern the evolution of the perturbations that we have imposed. An important simplification occurs when the perturbations are *small*, in particular when $T' \ll \Delta T$; as long as this remains valid, the term $\mathbf{u}' \cdot \nabla T'$, being the product of two small quantities, is very much less than $w' \Delta T / h$ and can be neglected in the last equation of (4.95). If that term is omitted, the set of equations has important characteristics: it is a set of coupled partial differential equations that is *linear* in the field variables primed (no squares or products), and all the coefficients are *constant* (independent of x, y, z, and t). Just as in the theory of ordinary linear differential equations with constant coefficients, the general solutions of such a set are exponential in form, so that we should be able to find solutions for each of the variables that are proportional to

$$\exp(nt) \exp i(kx + ly + mz). \qquad\qquad (4.96)$$

This choice is not arbitrary. In our perturbation, we are free to impose at the initial instant any spatial distribution of disturbance consistent with the boundary conditions, but by Fourier's theorem this can be represented quite generally as the superposition of spatially oscillatory terms, as in a Fourier series or a Fourier integral. There is consequently no loss of generality in this spatial representation of the disturbance – if any one of these Fourier modes is unstable, the disturbance will grow. The quantity n (which is to be determined) may be complex, but if its real part is negative for all (k, l, m), all disturbances will die away. If it is positive for at least some (k, l, m), those modes will grow exponentially in time until the neglected term $\mathbf{u}' \cdot \nabla T'$ in (4.95) becomes comparable with the others, and some new, more complex stable state may evolve.

Two final simplifications are useful. First, spatial periodicity in the horizontal direction represents an oscillation in the direction of the wave-number vector

$\mathbf{k} = (k, l)$. There is no preferred horizontal direction in the geometry of the problem, so that we can choose our x-axis in the direction of $\mathbf{k}$, causing the dependence on y to disappear and the problem to become two-dimensional. We can then use the stream function representation of Section 2.3 and eliminate the pressure from the first two equations of (4.95):

$$0 = -\frac{\mu}{k}\left\{\frac{\partial^2\psi'}{\partial x^2} + \frac{\partial^2\psi'}{\partial z^2}\right\} - \rho_0 g\alpha\frac{\partial T'}{\partial x}. \tag{4.97}$$

The linearized heat balance becomes

$$M\frac{\partial T'}{\partial t} + \frac{\partial\psi'}{\partial x}\frac{\Delta T}{h} = \kappa\nabla^2 T', \tag{4.98}$$

where Ψ' represents the stream function of the perturbed flow and (2.13) is used. The boundary conditions are now

$$T' = \frac{\partial\psi'}{\partial x} = 0 \text{ on } z = 0, h. \tag{4.99}$$

Second, to satisfy these boundary conditions without further ado and to use the separation h as a natural geometrical length scale, (4.96) can be particularized further by letting

$$T' = \hat{T}\sin\frac{m\pi z}{h}\exp(ilx/h)\exp(nt),$$
$$\psi' = \hat{\psi}\sin\frac{m\pi z}{h}\exp(ilx/h)\exp(nt), \tag{4.100}$$

where m is now a non-zero integer and the wave-number l is dimensionless.

The rest of the calculation is straightforward algebra. The substitution of (4.100) into (4.97) and (4.98) gives, after cancellation of terms, a pair of simultaneous homogeneous linear algebraic equations for $\hat{\psi}$ and $\hat{T}$. The condition for existence of a non-trivial solution (i.e. one other than $\hat{\psi} = 0$, $\hat{T} = 0$) is that the determinant of the coefficients vanishes, and this leads to a determination of the single quantity we want to know – the growth rate, n:

$$\frac{nh^2}{\kappa M} = \frac{l^2}{l^2 + m^2\pi^2}Ra - (l^2 + m^2\pi^2), \tag{4.101}$$

where

$$Ra = \frac{gkh\alpha(\Delta T)}{\nu\kappa}$$

is the Rayleigh number (4.21a).

It is apparent from (4.101) that in this context, the growth or decay rate n in the specification (4.96) is purely real. For stability the disturbance must die away,

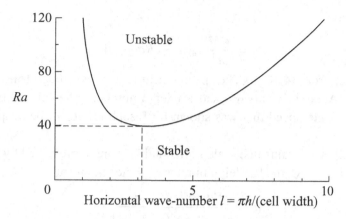

Figure 4.21. The Rayleigh stability curve for saturated porous media. At low Rayleigh numbers, the rest state is stable on all scales, but as the Rayleigh number increases to a value of 39.48, disturbances with horizontal wave-number π as indicated by the broken lines first becomes unstable. As the Rayleigh number increases beyond the critical value, so does the width of the band of unstable wave-numbers.

$n < 0$, and this requires that

$$Ra \leq \frac{(l^2 + m^2\pi^2)^2}{l^2} = f(l^2, m^2), \text{ say,} \qquad (4.102)$$

for all values of l and integral values of m. For any given choice of m, the function on the right of (4.102) increases as l^2 for large l and as $(m\pi)^4 l^{-2}$ for small l, and so has a minimum at some intermediate value, as seen in Figure 4.21. Points with coordinates (wave-number, Rayleigh number) above the curve specify unstable conditions and points below, stable. At the minimum

$$\frac{\partial}{\partial l} f(l^2, m^2) = 0;$$

this occurs when $l^2 = m^2\pi^2$, at which point the function f has the value $4m^2\pi^2$. This stability limit is most restrictive when $m = 1$, so that the critical Rayleigh number based on the horizontal permeability at which instability can first occur is

$$(Ra)_{\text{CRIT}} = 4\pi^2 = 39.48, \qquad (4.103)$$

as found by Lapwood (1948). Below this Rayleigh number, all infinitesimal disturbances are stable. The most unstable motion when Ra just exceeds its critical value is in the form of alternately rotating square roll cells whose width is equal to the depth of the layer. The speeds of the fluid motion in these first unstable modes at slightly supercritical Rayleigh numbers gradually increase with time at the rate exp(nt), where

$$n = \frac{\kappa M}{h^2} \{Ra - (Ra_{\text{CRIT}})\}, \tag{4.104}$$

as can be seen from (4.101). (The specific heat ratio M is defined following equation (4.92)). At slightly supercritical Rayleigh numbers, the cells ultimately stabilize with a finite speed that was shown by Joseph (1976) to be proportional to $\{Ra - Ra_{\text{CRIT}}\}$.

If the medium contains fine scale horizontal layering represented by permeabilities k_V and $k_H \gg k_V$, the Rayleigh number assumes the form

$$Ra^* = \frac{gh\alpha \, \Delta T}{\nu \kappa} \frac{k_H k_V}{\left(k_H^{1/2} + k_V^{1/2}\right)^2},$$

and the condition for stability is $Ra^* \leq \pi^2$, which reduces to (4.103) when the medium is isotropic. When $k_H > k_V$, interstitial fluid can move more freely in horizontal directions than in the vertical and, despite the fact that the buoyancy driving the instability acts vertically, the horizontal scale of the cells in this case is greater than the layer depth.

Experiments on the onset of Rayleigh–Darcy instability have been conducted by Elder (1967), Katto and Masuoka (1967), Combarnous and LeFur (1969), and by Murray and Chen (1989), all in isotropic media. Measured values of the critical Rayleigh number were generally within 20% of the value $4\pi^2$ predicted by the theory, the major uncertainties, as Murray and Chen pointed out, being in the precise determination of the permeability k. The convection cells that developed were in each case two-dimensional, and approximately square in cross section.

When the Rayleigh number is substantially supercritical, an increasingly wide band of wave-numbers becomes unstable. Notable contributions by Palm, Weber, and Kvernvold (1972), Rudraiah and Srimani (1980), Borkowska-Pawlak and Kordylewski (1985), and others have shown that in the isotropic case at least, the patterns of convection become less regular and much more complicated as newly unstable modes grow to larger amplitudes than the roll cells and supplant them.

4.7.2 A physical discussion

The basic physics of the Rayleigh–Darcy instability may be somewhat obscured by the algebraic details of the calculation above, but the essence of it can be understood more clearly from the structure of the physical balances involved. In this discussion, we will ignore numerical factors and coefficients of order unity in order to concentrate on the physical content of these balances to answer the question: why is there no sustained interstitial fluid motion until the temperature difference exceeds a critical value?

The primary physical statements are (i) the relation between the horizontal gradient of interstitial fluid buoyancy and the rotation vector of the transport velocity field. With flow in the (x, z) plane in the form of approximately square cells and the perturbation fields indicated by primes, the second component of (4.4) gives

$$-\frac{k}{\nu}\frac{\partial b'}{\partial x} = \Omega_Y \sim -\frac{\partial w'}{\partial x}$$

to within a factor of about 2, which we neglect. Thus

$$w' \sim \frac{k}{\nu}b' = \frac{gk}{\nu}\alpha T'. \tag{4.105}$$

The instantaneous vertical velocity distribution is directly proportional to the temperature perturbation from the undisturbed linear distribution.

(ii) The evolution of the two fields is specified by advection/diffusion balance in the field equation (4.98) for this temperature perturbation, which will be positive in some places, negative in others. Rewritten slightly, it becomes

$$\frac{\partial T'}{\partial t} = \frac{\Delta T}{h}w' + \kappa\nabla^2 T',$$

since the temperature gradient is negative upward. Multiply this last equation throughout by T' to form an equation for the *mean square* temperature perturbation, which is everywhere positive:

$$\frac{\partial T'^2}{\partial t} \sim \frac{\Delta T}{h}w'T' + \kappa T'\nabla^2 T',$$

and then form the average by integrating over the domain. Note that $T'\nabla^2 T' = \nabla\cdot(T'\nabla T') - (\nabla T')^2$. Either the perturbation temperature T' or temperature gradient $\partial T'/\partial n$ vanishes on the boundaries. Substitute from (4.105) and it follows that

$$\frac{\partial}{\partial t}\overline{T'^2} \sim \frac{\Delta T}{h}\frac{gk}{\nu}\alpha\overline{T'^2} - \kappa\overline{(\nabla T')^2},$$

$$\sim \frac{gk\alpha\Delta T}{h\nu}\overline{T'^2} - \frac{\kappa}{h^2}\overline{T'^2}. \tag{4.106}$$

This is an equation for the growth or decay of the (always positive) mean square temperature perturbation $\overline{T'^2}$ with purely numerical factors omitted. The first term on the right represents the rate of production of $\overline{T'^2}$ resulting from any buoyancy-driven vertical motion that distorts the basic constant gradient, and the second term represents the rate of dissipation of $\overline{T'^2}$ resulting from the smoothing effect of thermal diffusivity. The balance between them is independent of the magnitude of the perturbation $\overline{T'^2}$. The first term on the right is positive. If it is larger in magnitude than the second term, $\overline{T'^2}$ is increasing exponentially as a result of the distortion more rapidly than it is being smoothed out by the thermal diffusivity, the temperature perturbations will grow and the system is unstable. If the last term is larger in magnitude but negative, the sum of the two is negative and the mean square temperature perturbations die away exponentially as the system returns to rest. The ratio of the first term to the second is the Rayleigh number, whose numerical value at the point of balance, that is, of marginal instability, depends on the numerical constants that have been ignored.

4.7.3 Related configurations

The numerical value $4\pi^2$ for the critical Rayleigh–Darcy number in an isotropic medium is specific to the particular boundary conditions of constant temperatures and zero volume fluxes at the upper and lower surfaces. Nield (1968) found the critical values appropriate to a variety of other circumstances, showing, for example, that when the *heat flux* rather than the temperature is prescribed at upper and lower impermeable boundaries, the critical Rayleigh number is 17.7. Also, when the upper surface is at constant pressure (so that fluid can enter or leave across it) while the lower is impermeable, if the upper and lower temperatures are fixed, the critical value is 27.1.

When the porous layer slopes at an angle θ to the horizontal, the basic fluid state is not one at rest unless the isotherms happen to be horizontal. The particular case in which the temperatures at the upper and lower surfaces are uniform but with a temperature difference ΔT between them, has been studied extensively, both theoretically and experimentally (see, for example, Bories and Combarnous 1973, Combarnous and Bories 1975, and Caltagirone and Bories 1985). In this case, the isotherms also slope at the same angle, and in a uniform isotropic medium, the undisturbed flow is one of uniform shear with the rotation given by (4.5):

$$\Omega = -\frac{\alpha g k}{\nu}\frac{\partial T}{\partial x} = -\frac{\alpha g k \, \Delta T \, \sin\theta}{\nu h},$$

where h is the layer thickness. Various types of instability have been found to be possible, the simplest being longitudinal roll cells with their axes upslope. For these, the equations governing the transverse motions are identical to (4.97) and (4.98) (since $\partial/\partial y$, the derivative upslope, vanishes) except that $g\cos\theta$ replaces g. This type of instability then sets in when, for a uniform isotropic medium,

$$Ra\cos\theta \geq 4\pi^2, \tag{4.107}$$

as confirmed experimentally by Bories and Combarnous (1973). At much higher values of $Ra \cos \theta$, the flows become oscillatory and geometrically more complex; the various states and transitions among them having been considered in detail by Caltagirone and Bories (1985). When the layer is vertical and a constant temperature is maintained across it, the steady motion has uniform shear and is entirely stable, as shown by Gill (1969).

These analyses by Lapwood and later authors have been concerned with the problem of pure thermal convection, but the results can be extended immediately to the case of continuously saturated solutions in equilibrium with a major constituent of the fabric. The equation of state (2.65) replaces the last of the set (4.92); in the equations and therefore in the conclusions, the thermal density coefficient α_S of the continuously saturated solution replaces the ordinary thermal expansion coefficient α. For stability

$$\frac{g\alpha_S \Delta T h}{\nu k} \leq 4\pi^2.$$

Here, however, α_S may be negative, as it is for potassium salts and for Na_2SO_4 over a range of temperatures less than 32 °C, and if T is positive (heated at the bottom) the basic state is *unconditionally stable!* The increased density resulting from increased dissolution at the higher temperature overcomes the effect of thermal expansion. When $\alpha_S < 0$, however, the basic state can be destabilized by heating at the *top*, so that $\Delta T < 0$, $\alpha_S \Delta T > 0$, and again the critical value of above can be exceeded if the temperature difference is sufficiently large.

Another configuration of geological interest is that of a fluid layer of uniform depth above a saturated porous layer, heated from below. The stability characteristics of this system were studied theoretically by Falin Chen and C. F. Chen (1988) and experimentally in a series of measurements by the same authors in 1989. When the ratio $\hat{d}$ of the fluid layer depth to the porous layer thickness was close to zero, the upper surface was barely awash and the critical Rayleigh number was only slightly less than the value of 39.48 found by Lapwood. However, as the depth of the fluid layer was increased, the critical Rayleigh number (defined in terms of the porous layer parameters) showed a "precipitous decrease" to 21 when the depth ratio was 0.1. The physical reason for this appears to be the decrease in flow resistance in the fluid layer, which "shorts" the rising and falling fluid at the top of the porous medium cells. A further decrease to 2.71 at a depth ratio $\sim$0.2 was accompanied by an abrupt decrease in the critical wavelength by a factor of about 8 as the convective motion became more and more confined to the upper fluid layer, the porous medium Rayleigh number becoming less pertinent. Flow visualizations showed that the convection cells were generally three-dimensional. Figure 4.22,

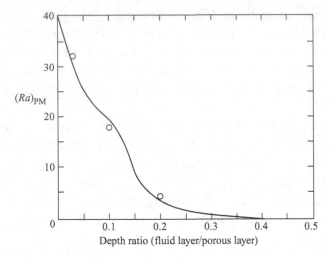

Figure 4.22. The critical Rayleigh number as a function of the depth ratio in a two-layer system with a liquid layer above a saturated porous medium. Measurements by Chen and Chen (1988 and (1989). Even a thin liquid layer short-cuts the circulation, reducing the flow resistance and reducing the critical Rayleigh number, defined in terms of the porous layer parameters.

re-drawn from the 1989 paper, compares results from their linear stability analysis with points from their measurements.

4.8 Thermo-haline circulations

In thermo-haline circulations, the local buoyancy of interstitial fluid is a function of both temperature and concentration of dissolved solutes, frequently salt, and these have different diffusion and advection characteristics in saturated permeable media. Recall that heat is conducted through the solid/fluid matrix with a thermal diffusivity (c.f. Section 2.7) $\kappa \sim 1.5 \times 10^{-3}$ cm^2/s. Heat is advected through the medium with an effective velocity $\mathbf{u}/M$, where $\mathbf{u}$ is the transport velocity and the specific heat ratio $M = (\rho C)_{\mathrm{M}}/(\rho C)_{\mathrm{F}} \sim 0.5$. The effective advective velocity is larger than $\mathbf{u}$ because of the high heat capacity of the moving water, but less than the mean interstitial fluid velocity $\overline{v}$ because of heat leakage into the matrix. If the interstitial fluid is not moving relative to the matrix, dissolved salts can diffuse through the matrix pores, but very slowly, because of three separate factors: (i), the porosity fraction $\phi < 1$, (ii) the tortuosity of the pathways, and (iii) the small molecular diffusivity of salt in water (c. 1.5×10^{-5} cm^2/s, smaller by a factor of about 100 than that for heat). If the interstitial fluid is moving through the medium, chemically passive, dissolved salts move through the medium at the mean interstitial fluid velocity $\overline{v}$ and are dispersed about the mean streamlines by mechanical (or geometrical) dispersion, though as Griffiths (1981) noted in his experiments, at the pore scale the mixing is initially local and sporadic until pore scale diffusion can smooth out the salinity variations. If there is significant interstitial fluid motion, the effective longitudinal diffusivity $D \sim \bar{v} \alpha_D$ (§§ 2.10, 3.3) may be very much larger than the molecular value and, in an accelerating flow, increases as the flow speed does.

In order to explore the interplay among the differences in advective velocities and diffusivities for heat and salt, it is instructive to reconsider the Rayleigh–Lapwood problem with a uniform temperature gradient $\partial T / \partial z = -\Delta T / h$ and a coexisting salinity gradient $\partial S / \partial z = -\Delta S / h$. This problem was first considered by Nield (1968) although, as Murray and Chen (1989) pointed out, his formulation unfortunately contained several slips. The equations describing small perturbations about a state of rest are the same as (4.97) and (4.98) with additional salinity terms

$$\frac{\partial^2 \psi'}{\partial x^2} + \frac{\partial^2 \psi'}{\partial y^2} = K \left(-\alpha \frac{\partial T'}{\partial x} + \beta \frac{\partial S'}{\partial x} \right),$$

$$M \frac{\partial T'}{\partial t} + \frac{\Delta T}{h} \frac{\partial \psi'}{\partial x} = \kappa \nabla^2 T', \qquad (4.108)$$

$$\phi \frac{\partial S'}{\partial t} + \frac{\Delta S}{h} \frac{\partial \psi'}{\partial x} = \phi \kappa_S \nabla^2 S',$$

where K is the hydraulic conductivity of the medium. It simplifies the algebra to use the physical characteristics of the system as scales to define dimensionless variables. We note that K has the physical dimensions of velocity and that the flow speeds are proportional to the buoyancy variations. Accordingly, let

$$S' = (\Delta S)\sigma, \qquad T' = (\Delta T)\theta, \qquad (x, z) = h(\xi, \eta),$$
$$\psi' = K\alpha(\Delta T)h\psi = (\kappa Ra)\psi, \qquad t = (h^2/\kappa)\tau, \qquad (4.109)$$

where the un-primed Greek symbols are dimensionless and Ra is the Rayleigh number (4.21a). In terms of these, (4.108) becomes

$$\text{generation of rotation:} \quad \frac{\partial^2 \psi}{\partial \xi^2} + \frac{\partial^2 \psi}{\partial \eta^2} = -\frac{\partial \theta}{\partial \xi} + R_\rho \frac{\partial \sigma}{\partial \xi},$$

$$\text{thermal energy balance:} \quad M \frac{\partial \theta}{\partial \tau} + (Ra) \frac{\partial \psi}{\partial \xi} = \nabla^2 \theta, \qquad (4.110)$$

$$\text{salt balance:} \quad \phi \frac{\partial \sigma}{\partial \tau} + (Ra) \frac{\partial \psi}{\partial \xi} = \frac{\phi \kappa_S}{\kappa} \nabla^2 \sigma,$$

where the density ratio (Turner, 1973)

$$R_\rho = \frac{\beta \Delta S}{\alpha \Delta T} = \frac{\beta \partial S / \partial z}{\alpha \partial T / \partial z}, \qquad (4.111)$$

and the Laplacian operators are in dimensionless space variables. It is interesting to note that in the scaled set (4.110), all the coefficients are generally of order unity except for the diffusion term in the salt balance in which $\kappa_S/\kappa \sim 10^{-2}$.

The simplest boundary conditions are that the temperature and salinity are fixed, and that the normal component of the velocity vanishes at the upper and lower boundaries of the flow. This implies that $\theta = 0$, $\sigma = 0$ and $\psi = 0$ when $\eta = 0$ and 1.

To examine the stability characteristics of this system, the procedure is the same as in Section 4.7, but the algebra is somewhat more extensive. Since the coefficients in (4.110) are independent of ξ, η, and τ, solutions representing a typical Fourier component of the disturbance field are of the form

$$\exp(n\tau)\sin(m\pi\eta)\exp(il\pi\xi).$$

Note that the vertical half-wavelength of the disturbance is $1/m$, and the horizontal wavelength is $2/l$, where l and m are positive integers. The temporal growth rate, to be determined, is n. Substitution into (4.110) now leads to a quadratic equation for the growth rate n:

$$an^2 + bn + c = 0, \tag{4.112}$$

where

$$a = M \sim 0.5,$$

$$b = \pi^2(l^2 + m^2) - \frac{l^2}{\pi^2(l^2 + m^2)}(1 - MR_\rho/\phi)Ra, \tag{4.113}$$

$$c = \pi^4(l^2 + \pi^2)^2\frac{\kappa_S}{\kappa} + l^2\frac{R_\rho}{\phi}Ra,$$

and ϕ is the porosity. A term $\kappa_S/\kappa \approx 10^{-2}$ has been neglected compared with unity at one point.

The growth rate for the disturbance is determined by the nature of the roots of the quadratic equation (4.112), i.e.

$$n = (2a)^{-1}\{-b \pm (b^2 - 4ac)^{1/2}\}. \tag{4.114}$$

These roots are not necessarily real and there are four different cases that must be considered separately, corresponding to the two-layer situation illustrated in Figure 4.23.

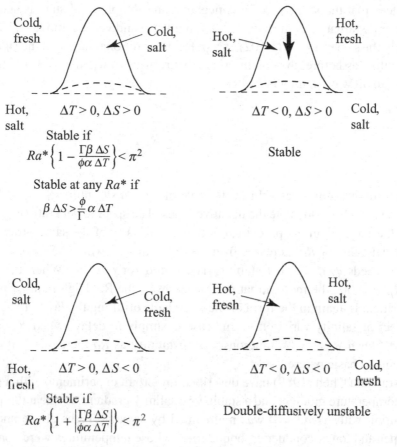

Figure 4.23. Cartoon of isohaline (solid lines) and isotherm displacements (dashed lines) in diffusive, saturated permeable media with the associated stability characteristics.

4.8.1 *Temperature destabilizing, salinity stabilizing,*
$$\partial T/\partial z < 0,\ \partial S/\partial z < 0,\ R_\rho > 0$$

In this case, cooler, fresher interstitial water lies above warmer, more saline water. When the salinity contrast is zero, it reduces to the Rayleigh–Lapwood problem of the previous section; one would expect intuitively that a distribution of salinity decreasing upward would enhance the stability, and this can be shown simply. The factors a and c are both positive and if $b^2 > 4ac$, the two roots are real and with sign opposite from b; if $b^2 < 4ac$, the roots are complex but the sign of the real part is still opposite to that of b. In either event, then, the condition $b > 0$, or

$$\pi^4(l^2 + m^2)^2/l^2 > \left(1 - \frac{M R_\rho}{\phi}\right) Ra$$

guarantees that the real part of n is negative and the physical state is stable to a particular disturbance of horizontal wave-number l and vertical mode m. This is of precisely the same form as (4.102) except for the additional factor on the right-hand side, so that, as before, the condition is most restrictive when $m = 1$ and the state is stable to all wave-numbers l when

$$Ra < 4\pi^2 \left(1 - \frac{M}{\phi} R_\rho\right)^{-1}. \tag{4.115}$$

The term in parentheses exhibits the role of the density ratio $R_\rho = \beta \Delta S / \alpha \Delta T$ in salinity stabilization. Note the negative index. The specific heat ratio is generally somewhat larger than the porosity, but they are usually of the same order, so that the critical value of Ra increases from its value of $4\pi^2$ when $\Delta S = 0$, to infinity as R_ρ exceeds ϕ/M, so that stability is ensured for *any* Ra. When the density ratio $R_\rho < \phi/M$, the motion that first develops as the Rayleigh number becomes supercritical is again in the form of the large cells of filling the layer. In summary, the effect of salinity variation in this case is simply to delay (when $R_\rho < \phi/M$) or avert (when $R_\rho > \phi/M$) the onset of instability as Ra increases, without any change in its basic nature.

Murray and Chen (1989) have described important experiments with a destabilizing temperature gradient and a stabilizing salinity gradient. A linear temperature distribution with $\partial T/\partial z < 0$ was maintained by separate baths above and below the upper and lower conducting boundaries, whose temperatures were controlled; the salinity gradient was produced by filling the box in four layers, which were allowed to diffuse. The internal salinity interfaces disappeared fairly rapidly, but since there was no salt flux across the upper and lower surfaces, the salinity gradient $\partial S/\partial z = 0$ there, and almost isohaline layers slowly developed and thickened gradually. Since the region of constant salinity gradient occupied only the central region of the depth interval, the result (4.115) cannot be applied directly, but in any event, the salinity stabilization effect was evident. With a strong salinity gradient, the critical Ra increased fourfold. When Ra was large enough to induce instability, large three-dimensional cells were formed and the heat flux increased dramatically. To account quantitatively for the measurements on the onset of instability, it was necessary to take into account both the nonlinear salinity profile and the temperature dependence of the thermal expansion coefficient – a reminder that the simple criteria such as (4.115) are important conceptually but probably unreliable numerically if applied to geological situations in which the conditions envisioned in the theory (both gradients constant) are not precisely satisfied.

4.8.2 Both temperature and salinity stabilizing,
$$\partial T/\partial z > 0, \, \partial S/\partial z < 0, \, R_\rho < 0$$

In this situation, hot, fresh interstitial water lies over cold, more saline water; in terms of the problem definition, $\alpha \Delta T < 0$, $\beta \Delta S > 0$. Consequently, $Ra < 0$ and $R_\rho < 0$. In (4.114), c is then positive for all l, m, as is b. The two roots (4.114) then have negative real parts, and the equilibrium is, not surprisingly, always stable.

4.8.3 Both temperature and salinity destabilizing,
$$\partial T/\partial z < 0, \, \partial S/\partial z > 0, \, R_\rho < 0$$

Now cold, more saline water lies over fresher, hotter water; $\alpha \Delta T > 0$ and $\beta \Delta S < 0$, so that $Ra > 0$ and $R_\rho < 0$. From (4.114) both roots are negative (or have negative real parts) if $c < 0$ and $b > 0$ for all l, m, and this can be shown again to require the condition (4.115). Since, however, $R_\rho < 0$ in this case, the factor in parentheses is larger than unity, so that the critical Rayleigh number decreases toward zero as the negative term MR_ρ/ϕ becomes larger in magnitude – diffusive smearing is less and less able to maintain stability.

4.8.4 Temperature stabilizing, salinity destabilizing,
$$\partial T/\partial z < 0, \, \partial S/\partial z > 0, \, R_\rho < 0$$

This situation, when hot, saline interstitial water lies over cooler, fresher water, is possibly the most interesting, being associated with the phenomenon of double-diffusive fingering. The latter is similar in its basic dynamics to the double-diffusive instability in ordinary viscous fluids described by Turner (1973) in his fine book *Buoyancy Effects in Fluids*. If an element of interstitial fluid is displaced downward, it loses heat by diffusion but hardly any salt; it is then denser than the surrounding fluid and tends to continue downward. If displaced upward, it becomes warmer than its surroundings but remains fresher, and so is less dense and continues upward. Interleaving fingers or vertical sheets then form and continue as long as the salt flux downward is maintained. This instability was predicted theoretically by Nield (1968), but has not yet been demonstrated in the laboratory in a porous medium, heat–salt system because of the experimental difficulties associated with lateral heat losses in an apparatus of limited horizontal dimensions. Imhoff and Green (1988) have shown, however, that this kind of fingering develops in the much less efficient double-diffusive system using gradients of sugar and salt in the interstitial fluid. Salt (NaCl) has a higher molecular diffusivity than does sugar, but by a factor of only 3, and the effective advection velocities are the same for the two species, whereas in the heat–salt system, the thermal diffusivity is greater than that for salt by a factor of 100 and the salt advection velocity is larger by the factor M/ϕ. Nevertheless, even in this less efficient case (the more diffusive species, salt, having the same dynamical influence as cold and the less diffusive one, sugar, playing the same role as salt in the heat–salt system), fingers developed that were about ten times as long as they were wide. The net vertical fluxes were about two orders of magnitude larger than the molecular diffusion in a motionless interstitial fluid. In a heat–salt system, the motion would be expected to be more vigorous and the fingers relatively longer in view of the greater diffusivity contrast together with the more rapid salt advection compared with that of heat.

Algebraically, one of the roots of (4.114) is real and positive (indicating instability) when $c < 0$. In the present situation, $\alpha \Delta T < 0$ and $\beta \Delta S < 0$, so that the density ratio R_ρ is positive and $Ra < 0$. For small-scale, vertical disturbances $m = 0$ and from (4.112),

$$c = l^2 \left\{ l^2 \frac{\kappa_S}{\kappa} + Ra \left(\frac{R_\rho}{\phi} - \frac{\kappa_S}{\kappa} \right) \right\}, \qquad (4.116)$$

and $c < 0$ when

$$l^2 \frac{\kappa_S}{\kappa} < |Ra| \left(\frac{R_\rho}{\phi} - \frac{\kappa_S}{\kappa} \right).$$

Accordingly, when

$$R_\rho = \frac{\beta \Delta S}{\alpha \Delta T} > \phi \frac{\kappa_S}{\kappa}, \qquad (4.117)$$

the basic state is unstable to some wave-numbers at any Rayleigh number $|Ra|$. The right-hand side of (4.117) is numerically very small. When $R_\rho < 1$, the density of the interstitial fluid decreases in the vertical and the basic state is apparently statically stable; however, it is unstable in this double-diffusive manner when

$$\phi \frac{\kappa_S}{\kappa} < R_\rho < 1.$$

In practical terms, since $\phi \kappa_S / \kappa$ is so small, unless the salinity gradient is vanishingly small, a situation of hotter, more saline interstitial fluid lying over cooler, fresher fluid is *always unstable* even if the overall fluid density decreases in the vertical direction. The resulting circulations are in the form of interleaving small vertical fingers or sheets, providing effective vertical transport of dissolved ions in the interstitial fluid. The case of cooler, more saline water overlying fresher, warmer water is also highly unstable, so that as a general rule, instability and vertical transport are almost inevitable when more saline interstitial water overlies a fresher interstitial fluid region, whatever the temperature structure. Geological scenarios of this kind are not at all uncommon, particularly in evaporite situations discussed in Chapter 5.

4.8.5 Brine invasion beneath hypersaline lagoons

When a coastal embayment in an arid climate becomes landlocked, the excess of evaporation over precipitation leads to an increasing salinity in the lagoon water, and it becomes hypersaline relative to a seawater-saturated fabric below. Even if the possibly hot lagoon water is less dense than that in the substrate, the stage is set for double-diffusive fingering of the interstitial water, leading to vertical solute transfer and geochemical alteration of the host rock. This may well have occurred in the formation of certain bedded dolomites by the post-depositional alteration of limestone shelves of the Permian Basin in the southwestern United States.

The origin of dolomites was regarded as enigmatic for many years. The principal problems, as listed by Adams and Rhodes (1960), included (i) the existence of an adequate source of chemically active magnesium to allow the degree of dolomitization observed, (ii) the existence of a plausible and natural internal flow system that can transport the magnesium into, and distribute it throughout, the host limestone, while at the same time removing the displaced calcium ions, and (iii) the provision of a chemically favorable environment in which the reaction can occur. Commonly, dolomite is associated with beds of evaporite such as salt, gypsum, or anhydrite. Often, however, there is no such association. Some coral atolls are extensively dolomitized at depth. This suggests that there is no unique physical or geological environment associated with dolomitization – the chemistry may be the same, but the internal flow systems needed for magnesium transport may be of several kinds, the circulation beneath hypersaline lagoons being only one.

The Permian Basin of western Texas and New Mexico is a huge area, rich in petroleum, natural gas, and potash, with excellent outcrops as a result of uplift and erosion. It has been extensively drilled. The economic importance of the region has stimulated many studies reconstructing the Permian geography. Some 250 millions years ago, this region was the site of a partly landlocked embayment extending more than 1500 km into the continental interior with shell- and skeleton-forming shallow-water fauna building the basin floor upward as subsidence lowered it. Adams and Rhodes (1960) visualized extensive shallow lagoons perhaps 300 km wide and 3 m deep, evaporating in an arid climate but sporadically replenished by new seawater, so that the salinity gradually increased from that of normal seawater (35‰) to values five or even ten times higher when halite (rock salt) deposits began to form. The hot, saturated solution with excessively high concentrations of magnesium, potassium, sodium, and chlorine provides the chemical source of dolomite, but in order to distribute the solutes through the carbonate substrate, Adams and Rhodes were forced to postulate a dynamically implausible, density-driven flow pattern (which they called "seepage refluxion"). This hypothesis is clearly unnecessary in the light of the results above. The substrate with hypersaline interstitial fluid at the top and less saline fluid below is unstable to double-diffusive fingering, whatever the vertical temperature profile. This provides a natural (and inevitable) mechanism for the vertical transport downward of magnesium and other solutes in the downward-moving fingers and the transport up of calcium ions released in the dolomite reaction in the upward-moving ones. *The brine that provides the chemical source also drives the flow*. These simultaneous counterfluxes offer a conceptual solution to the transport problem in this environment, but the question remains as to whether the magnitudes of the fluxes are sufficient to provide the magnesium needed in a reasonable time interval.

In convective instabilities, the flow pattern that develops is in essence that of the most rapidly growing unstable mode. In a permeable matrix, as the disturbance grows, the diffusivity for salt increases as macroscopic dispersion replaces molecular diffusivity, and this relaxes the driving force for double-diffusive convection; the instability evolves toward a steady circulation. In a uniform medium, the fingers are vertical, but in a natural geological environment, the flow, though on average vertical, follows conduits or higher-permeability layers and is considerably less regular. Nevertheless, an order-of-magnitude estimate of the vertical fluxes can be found simply from the first of equations (4.4), or by balancing the vertical driving force (the horizontal variation in buoyancy produced by salinity variations) with the resistance to flow in the medium:

$$\frac{\mu}{k}w = \rho_0 g \beta S'.$$

If ΔS represents the difference in salinity between the downward- and upward-moving fingers or sheets, the magnitude of the vertical velocities up or down is

$$w \approx \frac{k_V g \beta \Delta S}{\nu} = K\beta\Delta S, \tag{4.118}$$

where k_V is the permeability for vertical flow and K is the corresponding hydraulic conductivity (with the physical dimensions of velocity). The total volume of fluid entering (and leaving) the matrix per unit horizontal area per unit time is then also given by (4.118), and in a layer of thickness h, the volumetric replacement time, or the time required for a volume of water equal to the volume of the system itself to pass through the system is

$$T_V \approx \frac{h}{w} \approx \frac{h}{K\beta\Delta S}. \tag{4.119}$$

Numerical values are of interest. In the west Texas, New Mexico basin, the thickness $h \sim 500$ m, and K may be as small as 10^{-7} m/s. Now $\beta\Delta S$ represents the fractional difference in densities resulting from salinity differences in the downward- and upward-moving columns. This is of the same order but somewhat less than the density difference between the hypersaline lagoon water and the interstitial water deep in the bank. As time passes, both become more saline, but the flow continues as long as the difference remains. If, for example, we take $\beta \Delta S \sim 0.05$ and $\nu = 10^{-6}\,\mathrm{m^2/s}$, then

$$w \sim 5 \times 10^{-9}\,\mathrm{m/s} = 15\,\mathrm{cm/yr},$$

and

$$T_\mathrm{V} \sim 3000\ \mathrm{yr},$$

a very short time.

The time required to complete the conversion of limestone to dolomite is, of course, greater than this, since the molar concentration of magnesium in even saturated brine is less than that of the dolomite produced. An accurate calculation requires a detailed chemical model of the sort outlined by Wilson *et al.* (1990), but the thermodynamic parameters in this case are not yet well established. A lower limit for the time interval can be obtained by assuming that all of the magnesium in the fluid is available for conversion to dolomite. One liter of dolomite contains about 11 moles of magnesium, and if the concentration in the convecting fluid is about ten times that of normal seawater, as Adams and Rhodes (1960) suggest, its concentration is about 0.5 molal. Some twenty-two flushing times are thus required, giving a lower limit of about 70 000 years. Because of the expected inhomogeneity of the medium and internal channeling, the actual time for dolomitization is certainly much greater than this, but even if this estimate is too small by a factor of 10, it is still sufficiently high to provide the needed magnesium within a geologically short time.

Another example of apparent thermohaline overturn has been described by Hanor (1987). Preferential dissolution near the top of the Iberia salt dome, Louisiana, combined with bleeding of less-saline water from a geopressurized region below, provides salinity destabilization with hot, saline water over cooler, fresher water, as described in the preceding section. Hanor used the simple (incorrect) criterion of a decrease in density of pore fluids with depth to infer the existence of convective instability, but the results earlier in this section indicate that instability and convective circulation will occur more widely than he anticipated.

4.9 Instability of fronts

A quite different kind of instability, very important in the petroleum industry, occurs when one interstitial fluid of viscosity μ_1 and density ρ_1 is displaced by another of viscosity μ_2 and density ρ_2. Though its existence seems to have been known in the oil industry for some time, the first account of it was published in

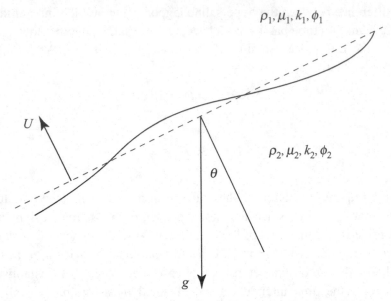

Figure 4.24. A definition sketch for a perturbed moving interface between saturated regions of differing density and viscosity.

the scientific literature by Hill in 1952. Saffman and Taylor (1958) gave a clear explanation of the basic mechanism so that the instability is generally associated with their names. Dynamically identical but geologically very different is the infiltration instability of dissolution fronts along which a minor constituent of the matrix is dissolving, producing a change in the matrix porosity and permeability across the front. These have been discussed extensively by Merino (1984), Ortoleva *et al.* (1987a), Ortoleva *et al.* (1987b), and their co-workers. To avoid repetition of the analyses, we also allow the permeability and porosity, as well as the viscosity and density to change across the front, allowing them to be equal or not, as appropriate to the geological application. In each application, capillary forces will be neglected for the moment, and the medium supposed to be locally isotropic.

   The nature of the instability can be seen most simply by supposing that the interface between the two fluid domains is sharp and initially plane, with its normal at the angle θ to the vertical. The interstitial fluid is moving through the matrix with uniform velocity W in the z-direction, normal to the plane. The transport velocity on either side is $U = \phi W$, also normal to the interface. Suppose now that the interface is not quite planar, but develops a small corrugation $\zeta = a \exp(imx + nt)$, as shown in Figure 4.24, with wavelength $2\pi/m$. The fluid is moving *into* the region with suffices "1". The disturbance grows in amplitude if n is positive and diminishes if negative. Since the perturbation is small ($am \ll 1$), the boundary moves through the medium with the interstitial velocity component on either side and the kinematical boundary condition (2.28) between the two regions simplifies to $\zeta = w_1 = w_2 = na \exp(imx + nt)$. The corresponding pressure distribution in front of and behind the interface satisfies Laplace's equation, and the disturbance pressures must vanish at infinity; with use of the condition above, they are found to be

$$p_1 = \rho_1 g z \cos \theta - \frac{\mu_1 \phi_1}{k_1} \left(Wz - \frac{na}{m} \exp(imx - mz + nt) \right), \qquad z > 0,$$

$$p_2 = -\rho_2 g z \cos \theta - \frac{\mu_2 \phi_2}{k_2} \left(Wz + \frac{na}{m} \exp(imx + mz + nt) \right), \qquad z < 0,$$

where $m > 0$ since the disturbances vanish far from the interface. These pressures must be the same on either side of the interface where $z = \zeta = a \exp(imx + nt)$, and when $am \ll 1$, it follows after some manipulation that

$$\frac{n}{m} \left(\frac{\mu_1 \phi_1}{k_1} + \frac{\mu_2 \phi_2}{k_2} \right) = (\rho_1 - \rho_2) g \cos \theta + \left(\frac{\mu_1 \phi_1}{k_1} - \frac{\mu_2 \phi_2}{k_2} \right) W. \quad (4.120)$$

A number of interesting and physically comprehensible special cases can readily be extracted from this equation. The term in brackets on the left and the spatial decay rate m are always positive. The growth rate n is real and the motion is unstable ($n > 0$) if the right-hand side is positive and stable if it is negative ($n < 0$). In the Saffman–Taylor instability, the porosity and permeability are the same on each side, so that $\phi_1 = \phi_2 = \phi$, $k_1 = k_2 = k$ and the question of stability or instability then depends only on the relative magnitudes of the densities and the viscosities. The condition for instability ($n > 0$) is then

$$(\rho_1 - \rho_2) g \cos \theta + (\mu_1 - \mu_2) \frac{\phi W}{k} > 0. \quad (4.121)$$

When the two viscosities are equal and $\cos \theta > 0$, the interface is unstable when the upper fluid is denser and stable otherwise, in accordance with intuition. Alternatively, when buoyancy forces are negligible, that is, when the fluid densities on each side of the interface are the same, or when the interface is vertical, the criterion for instability is, from (4.121), very simply that $\mu_1 > \mu_2$, that more viscous fluid is being displaced by less viscous fluid. In physical terms, this instability develops when a local indentation or finger of less dense fluid occurs in a moving interface, the pressure gradient normal to the interface forces the less viscous fluid inside the finger to move faster than the more dense fluid surrounding it, so that the finger grows.

More generally, when the fluid densities on either side differ, buoyancy effects may also be significant. When, again, $\mu_1 > \mu_2$, the interface is unstable when

$$W > -\frac{(\rho_1 - \rho_2)}{(\mu_1 - \mu_2)} \frac{gk \cos\theta}{\phi} \tag{4.122}$$

and this now depends on the orientation and speed of advance of the front, as well as the density and viscosity differences. For an interface moving upward, $\cos\theta > 0$. When the upper fluid is more viscous but less dense, $\mu_1 > \mu_2$ but $\rho_1 < \rho_2$, and the gravitational (buoyancy) effect promotes stability whereas the difference in viscosities promotes instability. The interface is then unstable when the speed of advance W is greater than a critical value,

$$W > W_C = \frac{(\rho_2 - \rho_1)gk \cos\theta}{(\mu_1 - \mu_2)\phi},$$

and stable when $W < W_C$. When the upper fluid is more dense as well as more viscous, both the gravitational and viscous effects promote instability; the right-hand side of (4.122) is negative and the interface always unstable. There are various other cases depending on the signs of $(\rho_1 - \rho_2)$, $(\mu_1 - \mu_2)$ and $\cos\theta$ which the reader can explore. A curious one is the viscous stabilization of a gravitationally unstable, downward-moving interface when heavy, more-viscous fluid above displaces lighter, less-viscous fluid below.

Dynamically identical but geologically very different is the infiltration instability of dissolution fronts that form when unsaturated interstitial fluid encounters a region containing soluble minerals which then dissolve. It will be seen in the next chapter that these fronts characteristically move much more slowly than the initially unsaturated interstitial fluid, which becomes saturated as it moves through the front. The dissolution process increases the medium permeability and porosity, but the interstitial fluid densities and viscosities are now virtually identical on either side of the front. In (4.120), the condition for instability of the frontal region, that $n > 0$, reduces to

$$\frac{\phi_1}{k_1} - \frac{\phi_2}{k_2} > 0.$$

For a particular medium geometry, the permeability $k \propto \phi^n$, where $n > 1$, the condition above is equivalent to $k_2 > k_1$ and $\phi_2 > \phi_1$. For infiltration instability, the permeability and porosity must be greater behind the front than ahead of it, and as the perturbations grow into fingers, flow continues to be focused into the advancing, more permeable intrusions, increasing the rate of dissolution. A precipitation front, or indeed any front in which the reaction reduces the permeability, is, in contrast, stable and perturbations of the frontal surface tend to disappear. A more detailed analysis of the stability of an acid dissolution front moving through porous rock has been given by Hinch and Bhatt (1990).

Some measurements in porous media have been made, notably by Slobold and Thomas (1963), using X-rays to observe the spreading and by Habermann (1960) using a relatively thin slice of permeable medium; they also show similar fingering phenomena. Observations are made difficult by the opacity of the medium, but Chouke, Van Meurs and van der Poel (1959) and Wooding and Morel-Seytoux (1976) describe observations with fluids having the same index of refraction as the medium, so that dye streaks in the interior become visible. Many numerical, random-walk simulations, following those by Sherwood (1986), have been interpreted in terms of fractals, presumably reflecting the absence of natural length scales in these processes.

5

Patterns of reaction with flow

5.1 Simple reaction types

A variety of chemical reactions can occur as water, carrying various dissolved chemical species, moves through a permeable matrix or a network of fissures or fractures. The *nature* of the resulting dissolution, precipitation, and fabric alteration depends on the reaction kinetics and the influence of temperature, pressure, and other factors on them. Their *spatial distribution* depends largely upon the flow. Compton and Unwin (1990) identify a series of about ten sequential steps that may be involved in any particular reaction. Of these, the controlling, or rate-limiting steps may be those in which reactants in solution are delivered from their source to the reaction site by advection and dispersion in the interstitial flow, and reaction products are carried away. Typical fluid velocities may be only 1 m/yr. In these flow-controlled reaction scenarios, there are two sets of balances to be considered: the chemical balances that operate at the molecular or ionic scale and define the nature of the reaction, and the physical balances that specify the spatial transport of dissolved reactant to the reaction site, and of fluid reaction product from it. In laboratory experiments and chemical engineering processes, the physical processes can be accelerated by stirring, but in many geochemical reactions in rock fabrics the natural transport processes can be so slow that they control the rate at which the overall reaction can proceed. Nevertheless, their detailed spatial distributions must reflect the chemistry, so that it is useful to review briefly a few of the simplest and most common chemical reaction types. In our discussion of chemical transport and reactions, it is convenient to express concentration in terms of molarity, moles per unit volume, a mole being a mass unit equal to the molecular weight of the species involved, so that the ratios of moles produced or consumed per unit volume are simple integral multiples. The flux of chemical species in physical space is given by the molarity times the fluid transport velocity through the medium. In precise geochemical calculations, the molality (moles of solute per unit mass of *water*) is a

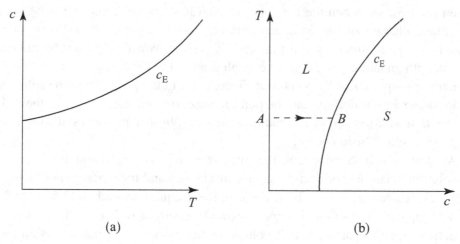

Figure 5.1. (a) A conventional solubility diagram and (b) the corresponding phase diagram, with the solid phase below and to the right of the curve c_E, and liquid solution to the left and above.

more convenient quantity, but unless the temperature and pressure are near critical, the difference is minor. A clear discussion of these matters is given in Garrels and Christ's classic book (1965).

5.1.1 Dissolution

The simplest type of chemical reaction is pure dissolution or precipitation, in which a solid mineral S dissociates reversibly into dissolved constituents or ions D_1 and D_2 though even this involves a sequence of perhaps ten intermediate chemical products. Overall, the reaction might be summarized as

$$S \leftrightarrows D_1 + D_2, \tag{5.1}$$

and is exemplified by the dissolution or precipitation of halite

$$NaCl \leftrightarrows Na^+ + Cl^-.$$

At equilibrium, where saturated solution and solid can coexist indefinitely, the saturation or equilibrium concentration c_{E1} of ion D_1 is a function of temperature, pressure, and also the concentration of ion D_2 – an *excess* of chloride ions drives the reaction above to the left and reduces the equilibrium concentration of sodium ions. With the other quantities held constant, the variation of saturation concentration with temperature can be represented by a solubility diagram, as in Figure 5.1a, or by an equivalent phase diagram, as in Figure 5.1b, in which c represents the molar concentration of one of the dissolved ions (say, D_1) in the interstitial fluid. At a given temperature, when the ion concentration lies along the curve $c = c_E$, solid and liquid phases can coexist in equilibrium; when $c < c_E$, the liquid (dissolved) state is in equilibrium only when no solid is present. Water may enter the soluble matrix with an initially negligible concentration of the ions involved, and its state is therefore represented by point A of Figure 5.1b. In the presence of the solute, it is no longer in equilibrium and the point representing its state moves to the right toward B, toward saturation, as simultaneously the physical processes of advection, dispersion and diffusion occur.

As described in Section 2.8, the rates at which the molecular processes of dissolution occur depends on the chemical kinetics and the surface morphology, and is expressed parametrically by the term Q_C in equations such as (2.51). On the flow scale, this is a function of temperature and pressure as well as (i) the proportion of active interstitial area per unit volume of the fabric or the mass fraction r of dissolving mineral present in the fabric, (ii) the relative concentration (X, Greek capital chi), the ratio of the concentration of solute to the equilibrium concentration c_E at the same temperature and pressure

$$X = c/c_E, \tag{5.2}$$

and (iii) the concentration of other solutes. In this definition, $c_E = c_E(T, p, c_1, c_2, \ldots)$ is the saturation or equilibrium concentration and c_1, c_2 and so on, represent the molar concentrations per unit volume of interstitial fluid, of the other dissolved species influencing the reaction (including possibly the hydrogen ion concentration, the pH). Thus, the source term per unit volume of fluid for either ionic constituent in the species balance equation can be written generally as

$$Q_C = \gamma c_E f(r, X), \tag{5.3}$$

where the "rate constant" γ has dimensions $[T^{-1}]$. In nature, the mass fraction r and relative concentration X both generally vary in space throughout the flow region and also possibly in time so that the source term Q_C varies correspondingly. The precise form of the dimensionless function f depends very much on the kind of reaction, on surface properties, etc and is not known in general.

In dissolution, the dissolution *rate* is a maximum when the solute concentration is very small and decreases as the solute concentration increases. Without loss of generality, we can take $\partial f/\partial X = -1$ when $X = 0$. This fixes the numerical value and the interpretation of γ in (5.3) as the reaction rate for dilute solutions. The dissolution rate is zero when the interstitial fluid is in equilibrium with the matrix or when there is no dissolving mineral left, so that $f(r, X) = 0$ when $X = 1$ or $r = 0$. In pure dissolution, then, the balance equation for either dissolved ion (assuming that it is not involved in another simultaneous reaction) is

$$\frac{\partial c}{\partial t} + \bar{\mathbf{v}} \cdot \nabla c - D \nabla^2 c = \gamma c_E f(r, X), \tag{5.4}$$

where $\bar{\mathbf{v}}$ is the mean interstitial fluid velocity and D the effective diffusion coefficient, the product of the dispersivity α_D (defined in Section 2.10) and the mean interstitial fluid *speed*, i.e. $D = \alpha_D \bar{v}$. In an isothermal situation, the saturation concentration c_E is often independent of position, and the previous equation then has the same form in terms of the relative concentration $X = c/c_E$ of (5.2),

$$\frac{\partial X}{\partial t} + \bar{\mathbf{v}} \cdot \nabla X - D \nabla^2 X = \gamma f(r, X). \tag{5.5}$$

An interesting asymmetry can be expected when precipitation occurs from a locally supersaturated solution ($X > 1$) since the *rate* of precipitation does not necessarily vanish when the mass fraction r of mineral already precipitated is small – a finite density of active nucleation sites for precipitation must be expected to exist even when $r = 0$, so that, in this case, $f(0, X) \neq 0$.

5.1.2 Combination

A second type of reaction involves the dissolution of two solid minerals S_1 and S_2, with the formation of a third solid mineral S_3 and a single dissolved species D:

$$S_1 + S_2 \leftrightarrows S_3 + D. \tag{5.6}$$

This is exemplified by the formation of wollastonite (calcium pyroxene) from calcite and silica in the sequence of reactions summarized by

$$\underset{\text{Calcite}}{CaCO_3} + \underset{\text{Silica}}{SiO_2} \leftrightarrows \underset{\text{Wollastonite}}{CaSiO_3} + CO_{2(Aq)}. \tag{5.7}$$

The silica may exist as impurities of quartz silt or chert in limestone, or the calcite may be an impurity in sandstone. In either case, the phase diagram can be represented in terms of the molar concentration of the dissolved solute (CO_2 in the example). At a given pressure and concentration of other ions influencing the reaction, the phase diagram for the reaction is as illustrated in Figure 5.2, where the equilibrium boundary $T = T_E(c)$ separates the (temperature–concentration) region below, in which $S_1 + S_2$ is stable, from that above, in which $S_3 + D$ is stable. Along the line $T = T_E(c)$ the four can coexist in equilibrium with aqueous interstitial fluid. The equilibrium is stable. At a fixed c a small increase in temperature above the line causes the balance in the reaction (5.6) to move to the right as long as both S_1 and S_2 are present, generating additional dissolved solute. The point representing

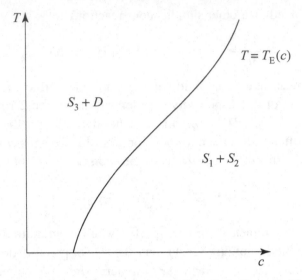

Figure 5.2. Phase diagram for the reaction $S_1 + S_2 \leftrightarrows S_3 + D$. The abscissa c represents the concentration of the dissolved solute D. All four phases can coexist in equilibrium along the curve $T = T_E(c)$.

the state of the system in Figure 5.2 then also moves to the right, back to the equilibrium curve $T = T_E(c)$.

Here again, the rate of generation of the dissolved species D can be represented by the expression (5.3), where r now represents the molar fraction of the *less abundant* mineral S_1 or S_2. If $r = 0$, only one of the two minerals on the left of (5.6) or (5.7) is present, the reaction cannot proceed, and no dissolved solute is generated. The spatial distribution of the reaction progress is monitored by the production and transport of the mobile solute D (carbon dioxide in the example above), whose distribution of relative concentration in space and time is specified by

$$\frac{\partial X}{\partial t} + \bar{\mathbf{v}} \cdot \nabla X - D\nabla^2 X = Q_C = \gamma c_E f(r, X). \tag{5.8}$$

Again, $f = 0$ when either the solution is saturated and $X = 1$ or when one of the initial minerals is exhausted and $r = 0$. Except for this circumstance, as each mole of the solute is generated in the reaction and carried away by the moving fluid, 1 mole of the mineral S_3 is being deposited and left behind. In (5.8), Q_C represents moles of solute generated per unit volume of the fluid, so that ϕQ_C is the number per unit volume of the *fabric*, and if s represents the moles of the immobile mineral S_3 (e.g. wollastonite) per unit volume of matrix, the rate of deposition of this mineral is

$$\frac{\partial s}{\partial t} = \phi n_S Q_C = \phi n_S \gamma c_E f(r, X). \tag{5.9}$$

The functions $f(r, X)$ in (5.8) and (5.9) are the same and n_S is the number of moles of the mineral deposited per mole of solute. We also have equations for the rate of disappearance of minerals 1 and 2 in the reaction. For the reaction-limiting, less-abundant species, with r moles per unit volume of fabric,

$$\frac{\partial r}{\partial t} = -\phi n_R Q_C = -\phi n_R \gamma c_E f(r, X), \tag{5.10}$$

where n_R is the number of moles of mineral dissolved per mole of solute generated. If the variation of r is significant, (5.10) and (5.8) are a pair of coupled equations for the concentration c (or, equivalently, the relative concentration X) of the fluid reaction product and the amount r of the less abundant solid species, whose solution in turn specifies the rate of deposition of S_3 by means of equation (5.9).

5.1.3 Replacement

A third type of reaction is replacement, in which aqueous ions D_1 entering the matrix replace another in the solid mineral S_1 to form S_2, as the new dissolved species D_2 is advected away. This can be represented as

$$S_1 + D_1 \leftrightarrows S_2 + D_2 \tag{5.11}$$

an example being the dolomite replacement reaction,

$$2CaCO_3 + Mg^{2+} \leftrightarrows CaMg(CO_3)_2 + Ca^{2+} \tag{5.12a}$$

or albitization of potassium feldspar,

$$KAlSi_3O_8 + Na^+ \leftrightarrows NaAlSi_3O_8 + K^+. \tag{5.12b}$$

The phase diagram for this type of reaction is illustrated in Figure 5.3, the abscissa representing the molar concentration of ion 2 as a fraction of the total molar concentration $c_1 + c_2$. When the concentration c_1 of entering fluid is zero, S_1 is stable at all temperatures since there is nothing to react with, so that the equilibrium curve slopes upward to the right. If, however, seawater containing Mg^{2+} but little Ca^{2+} (in the first example) enters a calcite bed, the reaction proceeds to the right of (5.12a), magnesium is lost from solution as calcium is added, and the reaction point moves horizontally to the right in Figure 5.3 until equilibrium is attained at that particular temperature. Again, as in the case of pure solution, the equilibrium concentration of solute D_1, i.e. c_{E1}, is a function not only of temperature and

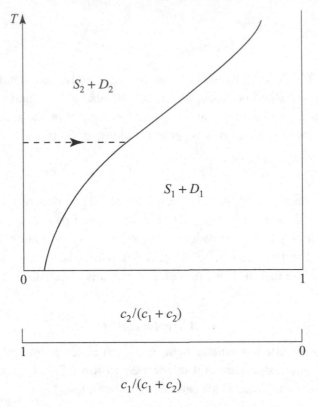

Figure 5.3. Phase diagram for the reaction $S_1 + D_1 \leftrightarrows S_2 + D_2$, where c_1 and c_2 represent the molar concentrations of D_1 and D_2 in solution.

pressure but also of the concentration of the product D_2. As before, the spatial distribution of reaction progress is monitored by the generation and spatial transport of the solutes involved in the reaction.

The flow equations for the dissolved species are similar to those already described, but now the source terms for the rates of change of the molar concentrations of solutes D_1 and D_2 in the interstitial fluid are equal and opposite. If the respective concentrations of magnesium and calcium ions are designated by c_1 and c_2,

$$\frac{\partial c_1}{\partial t} + \bar{\mathbf{v}} \cdot \nabla c_1 - D\nabla^2 c_1 = -Q_C,$$
$$\frac{\partial c_2}{\partial t} + \bar{\mathbf{v}} \cdot \nabla c_2 - D\nabla^2 c_2 = Q_C. \tag{5.13}$$

When the reaction proceeds to the right (as it does when magnesium-rich sea-water enters calcite), the sink of magnesium ions from solution is dependent on the surface density of mineral S_1 undergoing reaction and on the extent of disequilibrium, so that again

$$Q_C = \gamma n_1 c_{E1} f_1(r_1, X). \tag{5.14}$$

Since two molecules of $CaCO_3$ per molecule of solute are involved in the reaction (5.12a), $n_1 = 2$, and r_1 represents the number of moles per unit volume of S_1 as a fraction of the total solid moles per unit volume. X is the relative concentration (5.2) of D_1 (<1 if $c_1 < c_{E1}$). The rate of deposition of S_2 per unit volume of the matrix is again given by an equation such as (5.9).

In more complex reactions, similar balance equations can be written for the concentrations and transports of dissolved species, the main elaboration occurring in the specification of the equilibrium concentration c_E of each, which depends on the concentration of the others in the reaction. Note that in the balance equation for any particular species, the left-hand sides represent *rates of change* of concentration, so that the addition of an excess of any other species to the system simply affects the source term through the equilibrium concentration c_E.

When water is produced or consumed by the reaction, the rate of fluid generation produces a flow divergence,

$$\nabla \cdot \mathbf{u} = \phi n_W V Q_C, \tag{5.15}$$

where $\mathbf{u} = \phi \bar{\mathbf{v}}$ is the transport velocity, n_W is the number of moles of water generated per mole of solute, V is the fluid molar volume. When fluid is continually being driven through the matrix by either hydraulic or thermal forces, the incremental water addition may be slight, and in such cases, the modification to the flow velocity is insignificant, but there are important situations to be discussed later, involving dehydration reactions in only slightly permeable rocks, in which this represents the major water source.

In these expressions for the solute source terms Q_C, the reaction rates and the forms of the functions f for particular reactions are often not known to any degree of accuracy, but fortunately, even in this state of ignorance, useful results can be obtained, as demonstrated later in this chapter. Some limiting cases, however, do provide valuable simplification. When r is small, that is, when the reacting or *dissolving* mineral constitutes only a small fraction of the whole matrix, f can be expressed as the first (linear) term in a Taylor series expansion about $r = 0$ (for which $f = 0$), so that

$$f(r, X) \approx rg(X), \qquad \text{when } r \ll 1. \tag{5.16}$$

Note that $g = 0$ when $X = 1$, the solute being already saturated. Near equilibrium, again, when X is slightly smaller than one,

$$Q_c \approx \gamma c_E r(1 - X). \tag{5.17}$$

In *precipitation or deposition*, however, $f(r, X)$ does not generally vanish when the molar fraction r of the less abundant mineral is depleted, so that in this case

$$f(r, X) \approx f(0, X) = f(X), \text{ say.} \tag{5.18}$$

When r is small and if the solution is near equilibrium,

$$Q_C \approx -\gamma c_E (1 - X). \tag{5.19}$$

Finally, when r is large (nearly unity), the reacting solid constitutes almost the whole matrix, and the dependence of $f(r, X)$ on r can probably be neglected.

5.2 An outline of flow-controlled reaction scenarios

When solutes permeate through and react with a rock matrix, there are several characteristic physical scenarios that influence the spatial distributions of reaction, dissolution or deposition. The chemical reactions themselves may be the same under different physical scenarios, but the rates of mineral deposition and their spatial characteristics may be very different from one scenario to another. One of the main themes of this book involves the ways in which flow influences the rates of overall reaction, limiting them differently in different flow locations and at different evolution times in the history of a chemically active hydrological region. In nature, the growth of a sedimentary deposit and its chemical alteration may proceed simultaneously, or they may occur sequentially when, for example, the hydrothermal structure of an already existing region is modified in one way or another. For pedagogical rather than geological reasons, let us suppose that at some initial instant, a uniform stream of fluid containing reactive solutes enter the matrix across one of the boundary surfaces (or the water table) and continues to flow uniformly as the spatial pattern of geochemical alteration develops.

In this context, it is important to keep in mind the distinction between the (Eulerian) time derivatives at a fixed point and the (Lagrangian) time derivatives following the fluid motion, as specified in Section 2.3 (cf. equation (2.11)). The fact that a distribution of interstitial solute concentration remains steady in time ($\partial/\partial t = 0$), does not imply that there are no fluid–matrix reactions occurring. When the solute concentration varies spatially in a *steady* hydrological system, the solute load of individual fluid elements changes as they move through the fabric ($d/dt = \mathbf{u} \cdot \nabla$), indicating that fluid-matrix reactions are occurring. Consequently, the spatial pattern of interstitial solute concentration may not be changing as the chemical composition of the matrix alters. In this section, a brief survey is given of the several scenarios as an introduction to the more detailed discussion following.

5.2.1 The equilibration or reaction length

Consider an initially pristine water-saturated, permeable region of silica sand or carbonate sediment, say, and suppose that at some initial time, a uniform stream of chemically distinct water enters the region and as it percolates through, it begins to dissolve, or react with the solid matrix. The incoming fluid infiltrates into the region at the mean interstitial fluid velocity $\bar{v}$, characteristically a few decimeters or meters per year, and gradually tends toward a local equilibrium with the solid phase as Palciauskas and Domenico (1976) have pointed out in detail. If the kinetic rate constant defined in the previous section is γ, the interstitial fluid which is dissolving or reacting with the matrix, approaches equilibrium with it within a contact time of γ^{-1}, during which time the fluid elements have moved a distance

$$l_E = \bar{v}/\gamma. \tag{5.20}$$

This is called the equilibration length or the reaction length. Measurements of the overall rate constant γ, reported by Lerman (1979) and others, scatter widely over several decades about 1 yr^{-1}, so that field values of equilibration lengths may be of the order of meters or a few tens of meters. Near the boundary, the incoming fluid is far from saturation, the relative saturation is small (X $\ll$ 1 in equation (5.3)) and the rate of dissolution of the matrix or generation of reactant is greatest, but it decreases with distance from the interface as the interstitial fluid approaches saturation.

For definiteness, let us consider the arch-typical calcite–dolomite replacement reaction that occurs when sea-water, rich in magnesium, seeps through a calcite or limestone bed. The pattern of decreasing magnesium ion Mg^{2+} in solution is set up within the time that it takes for the individual fluid elements to move through the matrix a distance of two or three times the equilibration length, that is, over a time span of a few to a few tens of years. Then the distribution of concentration of the Ca^{2+} in solution produced in the reaction also stabilizes, being essentially zero at the entry point, increasing with increasing path length, and ultimately reaching equilibrium with the matrix after a travel distance of the order of the equilibration length. The solute distribution is then steady in time, though the reaction continues. Calcium ions derived from the reaction are carried downstream, and are continually augmented as the following fluid elements continue to interact and approach saturation. Once the fluid has moved beyond the equilibration length, it is close to equilibrium with the matrix and no further reaction or dissolution occurs.

Meanwhile, the initial solid reactant – the calcite CaCO$_3$ in the dolomite reaction (5.12a) – gradually becomes depleted and is replaced by dolomite, most rapidly near the interface where the incoming fluid is least saturated with respect to Ca^{2+}. This continues until most of the solid reactant accessible to the incoming, magnesium-rich fluid has already undergone reaction. At this stage, which occurs first near the entry interface or along permeable fractures that intersect it, the solid medium is primarily dolomite and unable to support further reaction, so that the incoming fluid begins to move unchanged past the interface and the reaction zone or front begins to move downstream.

5.2.2 The reaction front scenario

An order-of-magnitude estimate of the time needed to develop a separate front is given by considering the amount of water it takes to flush the products of reaction from the slice of the matrix adjacent to the boundary with thickness $l_E = \bar{v}/\gamma$ and unit cross-sectional area. If $2s_0$ is the number of moles of calcite per unit volume initially in the solid medium, then the number per unit area in the slice is

$$2s_0 l_E \approx 2s_0 \bar{v}/\gamma.$$

In accordance with the chemical balance (5.12a), half of the calcium has to be removed from the reaction site in the form of aqueous Ca^{2+} with concentration (number of moles of Ca^{2+} per unit mass of water) c_0. This requires a volume V of water per unit cross-sectional area such that $c_0 V = s_0 l_E$, so that $V = (s_0/c_0)l_E$, which is very much greater than the equilibration length. Since the transport velocity is $\phi \bar{v}$ where ϕ is the porosity, the requisite volume V of water per unit cross-sectional area is supplied in a time T such that $V = \phi \bar{v} T$. Consequently, the time needed for a front to form is

$$T \sim \frac{V}{\phi \bar{v}} = \frac{s_0}{c_0} \frac{l_E}{\phi \bar{v}} \sim \frac{s_0}{\phi \gamma c_0} \tag{5.21}$$

since $l_E = \bar{v}/\gamma$ as in (5.20). This is independent of the mean interstitial fluid velocity since the equilibration length is proportional to $\bar{v}$. Now s_0 is the number of moles of the solid reactant, calcite, per unit volume, and c_0 is the number of moles of calcite in aqueous solution, smaller by a factor of order 10^{-5}–10^{-6}. Thus $s_0/c_0 \sim 10^5$–10^6, the porosity $\phi \sim 0.2$–0.3 and the reaction rate $\gamma \sim 1$ yr^{-1} with much scatter. The time for the formation of a reaction front is then estimated from (5.21) to be about 10^5–10^6 yr, a huge multiple of the time it takes for the fluid elements to travel a distance equal to the equilibrium length. The reason for this is that the number of moles per unit volume of solid reactant initially in the matrix is so large compared with the number of moles per unit volume of even saturated aqueous solutions – were this not so, it would have leached out eons ago!

If the carbonate bed is extensively fractured, the incoming sea-water moves relatively rapidly along the fractures, as described in Section 3.6, reacting with the

$CaCO_2$ both along the fracture walls and also internally as it seeps into the interior of the matrix blocks. In hydrodynamic terms, the matrix blocks with higher overall porosity but lower permeability are the solute store-houses, the fractures with low fracture porosity and high permeability are the flow express-ways. Magnesium ions in solution move relatively slowly through each block where they are gradually exchanged with calcium ions at the kinetic reaction rate γ and are ultimately lost from the block by flushing back into the fracture network at the exchange rate E. Solute in the fractures moves, on average, much more rapidly and so is capable of traveling farther within the kinetic reaction time. On the other hand, since the overall volume fraction of fractures is small, with a given exchange rate, the mean residence time of fluid elements in fractures is much less than it is in the matrix blocks.

This description suggests that, in a fracture–matrix medium at a given distance into the reaction zone, the concentration of magnesium ions in the interstitial fluid in the blocks will be less than that in the fractures since the fluid elements in the blocks have had a longer time for the replacement reaction to proceed. We would anticipate on dimensional grounds that the thickness of the reaction zone, analogous to the equilibration length above, is proportional to the mean interstitial fluid velocity in the fractures (where the transport occurs) and inversely proportional to the reaction and exchange rates. This expectation is confirmed in Section 5.3 later.

The dissolution or reaction front that forms as described above, advances in the direction of the flow streamlines, but moves much more slowly than the interstitial fluid. The actual speed can be found from a simple mass balance (Figure 5.4) of the material entering the frontal region in solution where it undergoes reaction to form a solid product that remains as the moving front advances. The simplest case is exemplified by the dolomite reaction described chemically by (5.12a), namely,

$$2CaCO_3 + Mg^{2+} \leftrightarrows CaMg(CO_3)_2 + Ca^{2+} \qquad (5.12a)$$

In physical space, magnesium ions in saline solution are in equilibrium with dolomite as they move towards the reaction front between the dolomite and the calcite region. Here, the reaction above can proceed to the right, releasing calcium ions that flow away and forming the dolomite, which remains. The number of moles of magnesium per unit area per unit time that is incorporated as dolomite is $U s_0$, where s_0 is the moles per unit volume of magnesium in the solid mineral and U is the speed with which the front advances through the matrix. This is provided by the net influx of magnesium ions $\phi(c_0 - c_E)(\overline{v} - U)$ that move through the interstices with concentration c_0 as they catch up with the front and concentration c_E $(= 0)$ as

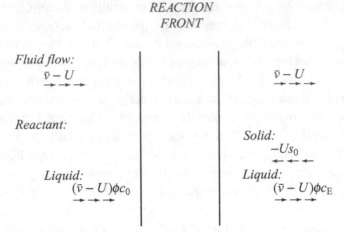

Figure 5.4. A flux diagram for simple leaching or dissolution of a solute in an inert permeable matrix. With interstitial fluid moving to the right with speed $\bar{v}$, fluid and mass balances determine the speed of advance, U of the dissolution front. In a frame of reference moving with this speed U, the flow pattern is steady, with the flux of solid undergoing dissolution is $U s_0$ to the left (towards the front) balancing the net flux of solute $\phi(\bar{v} - U)(c_E - c_0)$ to the right.

they leave it. Equating these two expressions leads to $\phi(c_0 - c_E)(\bar{v}_0 - U) = U s_0$,

$$\text{and so } U = \frac{\phi \bar{v}_0(c_0 - c_E)}{s_0 + \phi(c_0 - c_E)} \approx \frac{\phi \bar{v}(c_0 - c_E)}{s_0} \tag{5.22}$$

since $\phi c_0 \ll s_0$ by several orders of magnitude in these applications. Results equivalent to these have been given by by Lichtner (1985, 1991, 1992), Engesgaard and Kip (1992) and others, though not always correctly.

The ratio of molar concentrations appears again, but is now inverted, and $c_0/s_0 \sim 10^{-5}$–10^{-6} so that the propagation speed of the front is a tiny fraction of the mean interstitial fluid velocity. For example, if as assumed previously, the mean interstitial fluid velocity $\bar{v}_0$ is of the order of a few meters per year and the porosity is 0.2–0.3, then the speed of advance of the front is of order 10^{-5} m/yr. It would take a time interval of about 10^7 yr for the front to traverse a distance of 100 m.

5.2.3 The gradient reaction scenario

The term "gradient reaction" describes a different scenario. Imagine that in a water-saturated permeable region the matrix and the interstitial fluid are everywhere in local equilibrium. The equilibrium concentration of solute is generally a function of temperature and, to some extent, total pressure, and if the temperature and total pressure vary spatially throughout the region, so does the local concentration of dissolved species. If the interstitial fluid is at rest relative to the medium with local equilibrium between matrix and solute, no net dissolution, precipitation or other net reaction will occur. However, if the interstitial fluids begin to move through the matrix, the fluid elements find themselves in regions of different temperature and pressure, and for the fluid to move towards a new local equilibrium with the matrix, reactions must occur between the fluid and the surrounding matrix. For example, if interstitial fluid in a silica sand bed, initially in equilibrium with its surroundings, moves to a region of higher temperature, the fluid elements become unsaturated and silica must dissolve from the pore walls to restore equilibrium. If it moves to a lower temperature, silica is deposited on the walls. In general, the concentration of solute in each fluid element must change at a rate proportional to (i) the interstitial fluid velocity, (ii) the variation of equilibrium concentration with temperature (for example), and (iii) the spatial temperature gradient along the flow path. This is called the gradient reaction scenario, identified by Wood and Hewett (1982).

It has a number of particular characteristics that distinguish it from the passage of a reaction front in permeable-medium flow. Reaction occurs throughout the flow region simultaneously, but more rapidly along cracks or fractures that provide effective flow paths across isotherms. Gradient reactions can be expected particularly in geothermal regions where the temperature field may provide both the buoyancy distribution that drives the flow and the spatial temperature *gradient* that alters the equilibrium concentration along flow paths. Reactions proceed more slowly in relatively less permeable inclusions where the interstitial flow is reduced.

An analogy to the development of hidden internal reaction patterns in a permeable medium is provided by the very visible surface flow and reaction patterns in many geothermal areas. For example, the famous limestone terraces of Pamukkale, Western Turkey, are fed by a number of warm springs (about 35 °C) whose waters cascade through a series of fan-shaped pools and gradually cool as they spill from each level to the one below. The spring water contains many minerals in solution, including much calcium bicarbonate, which decomposes into calcium carbonate, carbon dioxide and water as the solution descends. The calcium carbonate is precipitated throughout the length of the cascade in the form of travertine (Ekmekci *et al.*, 1995, Altunel and Hancock, 1996), which has the appearance of gathered fabric. The rim of each spilling pool remains remarkably horizontal as the deposition proceeds, evidently stabilized by the increased flow and consequently increased deposition at any low points of the rim.

Just as the surface flow in this scenario continues to produce mineral alteration as long as the flow continues and the source minerals are not exhausted, so also does the interstitial fluid in a permeable-medium, gradient reaction scenario gradually modify the mineral composition, though less visibly. In a closed circulating flow, the processes of dissolution in some regions and deposition in others can redistribute minerals spatially along the flow paths as discussed in more detail in Section 5.5.

5.2.4 Mixing zones

Alterations in mineral composition or precipitation or dissolution can also occur by the mixing of different interstitial waters, although in the pores of a permeable medium, mixing at the molecular level required for chemical reaction is a much slower process than it is in a lake or in the sea. Mixing in rivers or the ocean is achieved by the almost-universal existence of a sequence of turbulent eddy sizes that strain the fluid elements on all scales, drawing them out into thinner filaments, increasing the concentration gradients and facilitating molecular diffusion. There is no analogue of this in a classical permeable medium, though the fracture–matrix structure does provide pathways for rapid flow and local mixing in the fractures. Simple mixing of water types inside a permeable medium appears to be a less universal phenomenon, occurring predominantly in a few particular flow situations, such as mixing zones in freshwater–saltwater interfaces, the vertical convergence of flow as in the aquifer flows described in Section 3.2, or by the focusing of flow into highly permeable lenses in which vertical gradients of solute concentration are amplified. When the equilibrium concentrations of a solute are different in fresh and saline waters, their mixture may produce dissolution or deposition, as described in Section 5.6.

5.3 Leaching or deposition of a mineral constituent

In the next few sections, these flow and reaction scenarios are described quantitatively in greater detail.

5.3.1 Dissolution in a uniform flow

The distribution of reaction (or dissolution or precipitation) in space and time depends not only on the solute balance equation (5.5), but also on the pattern of flow. The simplest case is presented by infiltrating water entering a permeable limestone bed across one of its bounding surfaces and moving uniformly through it, as illustrated schematically in Figure 5.5. As indicated in the previous section, the characteristic length scale of the reaction pattern is the equilibration length and if this is larger than the dispersivity of the medium, the advective solute transport by the mean flow generally dominates the random dispersion in the pores.

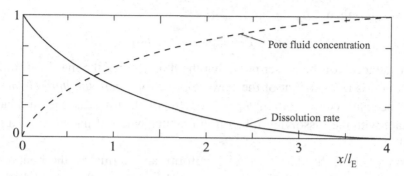

Figure 5.5. Initial distributions of solute concentration and dissolution rate when infiltrating fluid enters a permeable matrix with a soluble or reactive constituent. The pore fluid concentration increases and the dissolution rate decreases, on the scale of the equilibration or reaction length.

Equation (5.5) reduces to

$$\frac{dX}{dt} \equiv \frac{\partial X}{\partial t} + \bar{\mathbf{v}} \cdot \nabla X = \gamma f(X) \tag{5.23}$$

It is instructive to compare the Lagrangian and Eulerian descriptions of this simple situation. In a Lagrangian description, the mean distance that a fluid element has moved into the bed in time t is

$$\bar{x} = \int_0^t \bar{v}\,dt = \bar{v}t$$

from (2.68). The rate of change following the motion of the relative concentration of such a fluid element is given from (5.23) as

$$\frac{dX}{dt} = \gamma f(X), \tag{5.24}$$

where rate of reaction γ and the form of the function $f\,(0 < f(\mathbf{X}) < 1)$ are determined by the kinetics. In general, this equation cannot be solved explicitly in the form $X = X(t)$, but with a given $f(X)$, it can be integrated numerically. An implicit solution $t = t(X)$ with an initial concentration of zero is

$$t = \gamma^{-1} \int_0^X [f(X)]^{-1} dX.$$

Simpler and more informative explicit solutions can be found when the function f is linear in X. Since $f = 0$ when $X = 1$ (the solution being saturated), and $f = 1$ at $X = 0$ (the dissolution rate in very dilute solutions being simply γ), we take $f(X) = 1 - X$. The solution of (5.24) with the initial condition that when

$t = t_0$, $X = 0$ is then

$$X = 1 - \exp\{-\gamma(t - t_0)\}. \tag{5.25}$$

Note that, except for the prescription that the fluid entered the matrix at time zero, this solution is independent of the flow field, i.e. of *where the fluid element is* at time t. The solute concentration depends only upon the time that the fluid has been in contact with the dissolving mineral, no matter where it has gone in this time interval.

If, in particular, the flow is steady, uniform, and normal to the boundary, the expected distance the fluid element has traveled $x = \bar{v}t$ and the spatial distribution of interstitial fluid concentration is therefore

$$c/c_E = X = 1 - \exp(-\gamma x/\bar{v}) = 1 - \exp(-x/l_E). \tag{5.26}$$

where the equilibration length $l_E = \bar{v}/\gamma$. This solution can also be found (more simply) from the steady, one-dimensional Eulerian form of (5.23),

$$\bar{v}\frac{\partial X}{\partial x} = \gamma(1 - X), \tag{5.27}$$

as can be verified by substitution.

As anticipated in the previous section, *in a uniform flow*, the equilibration length is the characteristic distance from the incoming flow boundary over which the interstitial fluid is substantially under-saturated. It is the mean distance from the boundary that the fluid elements move in the reaction or dissolution time γ^{-1}. Because of the very great ranges in possible values of both the mean interstitial fluid velocity (of order 1 m/yr in un-fractured limestone to 100 m/yr in a sandy aquifer) and kinetic reaction rates, the equilibration length can assume almost any value in different formations; direct field observations of distributions of mineral alteration may provide the best measurement opportunities.

The solution (5.26) shows that in steady uniform flow, the rate of dissolution (mass per unit volume of the *medium* per unit time) at a point distant x from the entry surface, is constant in time and is distributed spatially as

$$\phi\bar{v}\frac{\partial c}{\partial x} = \phi\gamma c_E \exp(-\gamma x/\bar{v}) = \phi\gamma c_E \exp(-x/l_E), \tag{5.28}$$

where ϕ is the porosity. This is illustrated in Figure 5.5 also. When the distance from the incoming flow boundary is very small, $x \ll l_E$, the rate of dissolution is at its maximum, approximately $\phi\gamma c_E$, and is virtually independent of the flow speed. The dissolution process is here kinetically controlled. The dissolution rate (5.28) declines with increasing distance from the boundary as the solute concentration in the interstitial fluid increases, so that when $x = l_E = \bar{v}/\gamma$, the dissolution rate is only about 0.37 of that at the boundary and the solute concentration has risen to about 0.63 of its saturation value c_E. Since in a given mineral with a given kinetic reaction rate γ, the dissolution length scale or the equilibration length $l_E \propto \bar{v}$, and the rate of dissolution at an interior point depends strongly on the mean interstitial fluid velocity $\bar{v}$, the process has become largely *flow controlled*. If the flow regime continues over distances large compared with l_E, the interstitial fluid concentration in the interior approaches its equilibrium or saturated value, $c \rightarrow c_E$, and the rate of dissolution is choked off, becoming exponentially small. The total rate of dissolution is the integral of (5.28) from the entry point to the discharge. When the flow path length through the dissolving mineral is much longer than the equilibration length,

$$\text{total rate of dissolution} = \phi \gamma c_E \int_0^\infty \exp(-x/l_E)dx,$$
$$= \phi \gamma c_E l_E = \phi \bar{v} c_E = u c_E, \qquad (5.29)$$

where u is the transport velocity. This result is *independent* of the kinetic reaction rate γ, but is proportional to the flow velocity; it provides an overall check on the calculation. Water enters the region solute-free and leaves it saturated, so that the solute transport out is $u c_E$, as given by (5.29).

5.3.2 Leaching in aquifer flow with infiltration across the water table

The particular spatial distribution of dissolution (5.28) depends in detail on the assumption that f is a linear function of relative concentration X, but it depends far more on the uniformity of the pattern of flow. A more interesting kind of flow is exemplified by the two- and three-dimensional aquifer flows discussed in Section 3.2. Rainwater, infiltrating more or less uniformly across the upper surface $z = 0$, seeps vertically downward and at the same time spreads horizontally over distances proportional to the aquifer length l that is usually much greater than the effective flow depth d, and the patterns of dissolution or leaching are quite different from those given above. An important characteristic of these flows, established in Section 3.2, is that in an extensive aquifer with $l \gg d$, the distribution in dimensionless depth of the water "age," the time interval since the infiltrating fluid entered the region, is proportional to the recharge time but is independent of the plan form, or geographical shape of the flow region.

$$\tau_A = -T_{RC} \ln(1 + z/d), \qquad (5.30)$$

where the recharge time $T_{RC} = \phi d / \overline{W}$ as in (3.38), the basement is at $z = -d$ and the upper surface at zero. On average, the infiltration is equal to the discharge, so that the inverse of the recharge time can be interpreted as the flushing rate γ_F. The equilibration length l_E is the distance fluid elements have moved downward into the medium in the reaction time $\tau_A = \gamma^{-1}$. By inverting of the last equation, we then have

$$\frac{l_E}{d} = 1 - \exp\{-\gamma_F/\gamma\}. \tag{5.31}$$

The general nature of the distributions of interstitial solute concentrations can be inferred without further calculation. In a relatively rapid reaction, $\gamma \gg \gamma_F$ the equilibration length is small compared to the depth to the basement, so that below the infiltration surface the concentration of dissolving solute increases rapidly and reaches equilibrium at a relatively shallow depth. With a slower reaction in which the equilibration length may be of the same order as, or larger than, the depth, one would expect that the unsaturated region would extend more deeply, but a new consideration emerges. As is evident from (3.26) and (3.27) above, the mean interstitial velocity downward *decreases* with depth – the age of the infiltrating water becomes logarithmically large near the basement, and even though the reaction is slower, there is more time for equilibrium to be attained. One would then expect that under these conditions, most of the interstitial fluid in the upper part of the aquifer would be unsaturated with respect to the matrix where it is moving downward relatively quickly, while most of the more highly saturated fluid would be found in a relatively thin layer close to the aquifer basement.

Some simple calculations confirm these expectations. The vertical distribution of interstitial fluid concentration is found by eliminating $(t - t_0)$ from (5.25) and (5.30); after a little calculation, one finds that

$$X(z) = 1 - \exp\left\{\frac{\gamma}{\gamma_F} \ln\left(\frac{z}{d}\right)\right\}, \tag{5.32}$$

for $-d < z < 0$. This solution does satisfy the required boundary conditions since, at the upper surface $z = d$, the logarithmic factor is zero, $\exp(0) = 1$ and the relative saturation of the infiltrating fluid is zero, while near the basement as $z \to 0$, the logarithmic factor is large but negative and the exponential term vanishes. The interstitial liquid at the basement has been in contact with the medium for a long time and is virtually saturated so that, whatever the value of the ratio $R = \gamma/\gamma_F$ of reaction rate to flushing rate, the relative concentration $X \to 1$ there. An alternative form of the vertical distribution X can be given with use of the formula from elementary calculus, $a^b = \exp(b \cdot \ln a)$:

$$X(z) = 1 - (z/d)^R. \tag{5.32a}$$

This form of the solution satisfies the boundary conditions somewhat more obviously. The vertical distributions of rates of mineral alteration or dissolution under steady-state conditions can be found from these expressions, and they are illustrated in Figure 5.6 as functions of height above the basement for various values

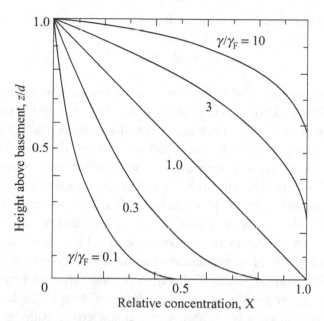

Figure 5.6. Vertical distributions of interstitial fluid concentration for flow reacting with the matrix in an aquifer with thickness d, porosity ϕ, mean infiltration rate W. The distributions depend on the ratio of the kinetic reaction rate γ to the aquifer flushing rate $\gamma_F = W/\phi d$. When the reaction rate is rapid, and the infiltration rate is small, γ/γ_F is large and the fluid reaches equilibrium with the matrix at a relatively shallow depth, and vice versa.

of R. They confirm the distributions of interstitial solute concentrations anticipated above.

When the dimensionless reaction rate $f(X)$ is linear in the relative concentration X, the distributions of dissolution rate are in essence the mirror images of the distributions of solute concentration. In a steady state with $c = c(z)$, the distribution of dissolution rate per unit volume of matrix is given by (5.3) as

$$Q_D = \phi\gamma c_E f(r, X) = \phi\bar{v}_z \frac{\partial c}{\partial z}.$$ (5.33)

In an aquifer, the distribution of vertical transport velocity is given in Section 3.2 as

$$\phi\bar{v}_z = -W(z/d),$$

while from differentiation of (5.32a),

$$\frac{\partial c}{\partial z} = -\frac{c_E \gamma}{d\gamma_F}\left(\frac{z}{d}\right)^{(\gamma/\gamma_F)-1}$$

so that the rate of mineral alteration or dissolution per unit matrix volume is

$$Q_D = \gamma \phi c_E (z/d)^{\gamma/\gamma_F}. \qquad (5.34)$$

This can be compared with the family of solutions representing concentration distributions (5.32a). When $\gamma/\gamma_F \gg 1$, $l_E/d \ll 1$ (rapid reaction kinetics, deep aquifer basement, slow flushing), most of the mineral alteration occurs in a relatively shallow, kinetically controlled layer as the downward-moving fluid interacts with the matrix and approaches a new equilibrium as shown by the top right curve of Figure 5.6. The rock alteration rate is small and the interstitial fluid is approximately uniform in a relatively thick layer above the basement. On the other hand, when $\gamma/\gamma_F \ll 1$ (slow reaction rates and rapid flushing), interstitial fluid is carried downward and the reactions have little time to move to completion until the fluid approaches the basement and its vertical descent slows. In this limit, the zone of mineral alteration or dissolution is limited to a relatively shallow region just above the basement, as shown in the lowest curve of Figure 5.6. The boundary between the two characteristic patterns occurs when $\gamma/\gamma_F = 1$, along which, according to (5.32a), the relative concentration $X = c/c_E$ increases linearly from zero at the water table at $z = 0$ to 1 at the basement, $z = -d$. The rate of dissolution decreases linearly with depth in the complementary way.

Numerical values are of some interest. Take $\gamma/\gamma_F = 1$ as the boundary between the two reaction patterns; of the other factors, assume as representative values an infiltration rate $W \sim 1$ m/yr, and $\phi \sim 0.2$, give or take a factor of 2 in each. The thicknesses of aquifer formations are quite variable. If it is taken to be 500 m, we find that kinetic reaction rates that are faster than 10^{-2} yr^{-1} have $l_E/d < 1$ and a relatively thin, under-saturated, kinetically controlled surface layer, beneath which the interstitial fluid is close to equilibrium with the dissolving mineral. With kinetic reaction rates slower than this, $l_E/d \sim 1$ and the corresponding surface layer is deeper. For a 100 m depth, the boundary occurs at 5×10^{-2} yr^{-1}. Values of kinetic reaction rates measured in the laboratory and cited in Section 2.8 (Lerman, 1979, Compton and Unwin, 1990) are much larger than these, being of order $(2 - 20)$ yr^{-1}. However, the laboratory and field situations are probably not comparable. In the laboratory studies, the dissolution surfaces are cleaned carefully to ensure repeatability, while in the field under natural conditions, the surface characteristics are unknown, but almost sure to be unclean with consequently slower kinetic reaction rates. Also, we must remind ourselves that natural aquifers are structurally variable (c.f. Section 3.3), so that these patterns must be interpreted as averages about which individual profiles scatter.

5.3.3 Dissolution in a fracture–matrix medium

As described in Section 3.5, freshwater entering a fracture–matrix medium moves relatively rapidly through the network of pathways in fracture planes, dissolving solute directly from the surfaces of the matrix blocks on either side. Water in the matrix blocks are also acquiring solute, presumably at the same rate, by dissolution inside the permeable blocks, but it is moving much more slowly, so that a given distance from the interface, the fluid in the matrix blocks has become more concentrated than that in the adjacent fracture network. At the same time, under the influence of the overall pressure gradient, less concentrated solution is entering the matrix blocks as more concentrated solution leaves them to rejoin the fracture flow.

The solute balances involved are essentially the same as those of Section 3.6, with additional source terms expressing the dissolution, which we assume to be linear in the degree of under-saturation, $(1 - X)$. Inside any individual matrix block, solute dispersion is small compared with advection, and by extension of (3.87), the solute balance per unit volume of fabric is

$$\phi_M \frac{\partial c_M}{\partial t} + \mathbf{u}_M \cdot \nabla c_M = \phi_M \gamma (c_E - c_M), \tag{5.35}$$

where c_E is the equilibrium or saturation concentration of the fluid in the interstices of the block and γ is still the kinetic reaction rate. As in Section 3.6, the volume integral of (5.35) over the block can be written as

$$\phi_M \int (\partial c_M / \partial t) dV + \int c_M \mathbf{u} \cdot d\mathbf{S} = \phi_M \gamma \int (c_E - c_M) dV,$$

with use of the incompressibility condition (2.8) and the divergence theorem. The first two terms can be represented as in equation (3.108) and, in the same spirit, the last term expressing the rate of addition of solute to the fluid in the block by dissolution per unit volume of the block can be approximated as $\phi_M \gamma (c_E - c_M)$. If we now average over all the blocks in our resolution volume, the solute balance in the matrix blocks reduces to (3.108) with an additional dissolution term.

$$\phi_M \frac{\partial c_M}{\partial t} = -\phi_M E (c_M - c_F) + \phi_M \gamma (c_E - c_M), \tag{5.36}$$

where $E = \bar{v}_M / l$ is the exchange rate, the inverse of the mean time for fluid elements to move through blocks of typical size l, and γ is the kinetic reaction rate. In terms of the relative saturation $X = c / c_E$, this last equation is

$$\frac{\partial X_M}{\partial t} = -E (X_M - X_F) + \gamma (1 - X_M).$$

By the time the entering fluid has moved through a distance of the order of the equilibration length (whose form is still unknown), the mean solute concentration in the fabric has become steady in time and the first term vanishes. The balance between production of solute by dissolution in any block is, in the steady state, balanced by the advective flushing expressed in the last two terms, so that the sum of the terms on the right vanish. A rearrangement of them gives

$$X_M - X_F = \frac{\gamma}{E + \gamma}(1 - X_F). \tag{5.37}$$

Note that the relative saturation in the matrix blocks, X_M, is greater than that in the fractures, X_F, although the difference diminishes with increasing distance into the medium.

In the solute exchange between fractures and matrix blocks, the rate of increase of solute in the fractures per unit volume of the fabric is equal to the rate of decrease in the matrix blocks, given by the first term on the right of (5.36), so that

$$\phi_F \frac{\partial c_F}{\partial t} + \phi_F \overline{v}_F \cdot \nabla c_F - \phi_F D \nabla^2 c_F = \phi_M E(c_M - c_F) + \phi_F \gamma(c_E - c_F).$$

In terms of the relative saturation, this is

$$\frac{\partial X_F}{\partial t} + \overline{v}_F \cdot \nabla X_F - D \nabla^2 X_F = \frac{\phi_M}{\phi_F} E(X_M - X_F) + \gamma(1 - X_F),$$

$$= \gamma \left\{ \frac{\phi_M}{\phi_F} \frac{E}{E + \gamma} + 1 \right\} (1 - X_F), \tag{5.38}$$

with use of (5.37).

As we saw earlier in this section, the flow controls the spatial distributions of reaction and dissolution rate. With unidirectional mean flow into the domain, the distributions of solute and dissolution rate become independent of time and equation (5.38) reduces to

$$\frac{\partial X_F}{\partial x} - \alpha_D \frac{\partial^2 X_F}{\partial x^2} = \frac{\gamma}{\overline{v}_F} \left\{ \frac{\phi_M}{\phi_F} \frac{E}{E + \gamma} + 1 \right\} (1 - X_F) = C \frac{\gamma}{\overline{v}_F}(1 - X_F), \text{ say,} \tag{5.39}$$

where α_D is the dispersivity and C is the factor in the large curly brackets whose significance will be explored shortly. When the fracture relative solute concentration X_F varies only on a scale much larger than α_D as would be expected with diffuse dissolution regions, the net effect of the dispersivity is small and the solution for the distribution of solvent concentration in the fractures becomes

$$c_F/c_E = X_F = 1 - \exp(-C\gamma x/\overline{v}_F), \tag{5.40}$$

so that in this context, the equilibration distance is

$$l_E = \overline{v}_F/C\gamma, \tag{5.41}$$

and the rate of dissolution per unit volume, concentrated in the fracture network, is

$$\bar{v}_F \frac{\partial c_F}{\partial x} = \gamma c_E C \exp(-C\gamma x/\bar{v}_F).$$

The form of these solutions is the same as the corresponding solutions for a "sandbank" medium (see equation (5.26)) except for the large factor C, in the curly brackets of (5.39). One would expect the fluid exchange rate in typical blocks, E, may be somewhat larger than the kinetic reaction rate γ for minerals like calcite, but the dominant factor contained in C is the porosity ratio $\phi_M/\phi_F \sim 10^3$. The equilibration length scale is decreased in a fracture–matrix medium about a thousand fold when compared with a "sandbank" medium, and the effective reaction rate is increased by the same factor. The physical reason for this effect is that most of the fluid is in the interstices of the matrix blocks, it moves slowly and has time to move toward equilibrium with the host rock before it has traveled far. Direct dissolution from the fracture network surfaces, represented by the last term in C, is expected to be negligible by comparison, although micro-cracks in the matrix blocks may convey fresher fluid some distance into the fracture walls. The mean concentration in the matrix blocks is, from (5.37), always somewhat larger than in the adjacent fracture pathways, since the solute concentration continues to increase because of dissolution during its sojourn there. Water in the fractures is a little fresher, but occupies a tiny fraction of the total void space, and moves into the fabric relatively rapidly, constantly acquiring and transporting solute from the surrounding blocks.

For aquifer flow with infiltration from above, the equilibration length is again smaller in a fracture–matrix medium by a factor of order ϕ_M/ϕ_F than it is in a "sandbank" medium with the same kinetic reaction rate, so that the ratio l_E/d governing the vertical distribution of dissolution is almost certainly much less than one in most aquifers. This suggests that, as a general rule, the interstitial fluid in a fractured calcite aquifer is generally close to saturation except for a thin surface layer, where most of the dissolution takes place.

5.3.4 The depletion time

If the dissolving mineral is contained in an inert matrix, the number s of moles per unit volume of the fabric decreases in the same manner as described by equation (5.10):

$$\frac{\partial s}{\partial t} = -\phi n_S \gamma c_E f(s, X), \tag{5.42}$$

where n_S is the number of moles of mineral dissolved per mole of solute produced. The distribution function $f(s, X)$ is of order unity when $X = c/c_E \ll 1$ (when the solution is not close to saturation) and s is not near zero (when the dissolving material would be depleted). If T_D represents the depletion time, the time derivative is of order s_0/T_D, so that from equation (5.42),

$$T_D \sim (\phi n_S \gamma)^{-1} \frac{s_0}{c_E}. \tag{5.43}$$

The initial solid molar fraction is generally larger than the molar interstitial fluid equilibrium concentration, c_E, by many powers of 10, so that these distributions persist for times enormously longer than the reaction time γ^{-1}. When, finally, the original mineral does approach depletion in the region where the fluid is entering, a reaction front develops, as described in the next section.

Parallel multiple reactions may involve more than one equilibration length. Those that are more rapid produce concentration distributions that change rapidly near the surface, whereas slower reactions have an influence distributed over greater depths. Examples of such profiles are given by Toth and Lerman (1977) and Lerman (1979).

5.4 The isothermal reaction front scenario

The reaction front scenario provides a physical setting that allows the chemical kinetics to transform large volumes of one mineral to another if it has a large enough source of fluid reactant, a mechanism to drive fluid through the matrix, and if it persists for a long enough time. Characteristically, calcite was deposited in shallow seas as shells over tens or hundreds of million years. Some magnesium may have been incorporated into the deposit at the time of its formation, but not nearly enough to account for the proportion of dolomite found in the deposits extant today. Continuing circulation with the "right" chemistry did not occur in all calcite deposits. Nevertheless, it is conceivable that, with the accumulation of the deposit, the geothermal temperature field inside may have been distorted in such a way as to generate internal convection patterns of flow. In the case of the Latemar Massif in Northern Italy, there is evidence discussed by Wilson *et al.* (1990), to suggest that this has happened, that the onset and continuation of local volcanic activity drove internal convective motions, circulating sea-water through advancing reaction fronts over a substantial length of time.

The nature of the mineral alteration and of the reaction products are consequences of the reaction kinetics, but certain overall properties such as the speed and direction of front propagation, the spatial patterns of mineral alteration in layered and fracture–matrix media and the fluid–rock ratios are *independent of the reaction kinetics* and can be found in a variety of chemical contexts. Three physical balances are involved – water mass conservation, the dissolved species balance, and an expression of the rate of deposition or dissolution of solid reactant.

In some reactions, water is produced or consumed, so that the incompressibility condition is (5.15), which we write in terms of the mean interstitial velocity $\bar{\mathbf{v}}$:

$$\nabla \cdot \bar{\mathbf{v}} = n_W V Q_C, \tag{5.44}$$

where Q_C is the rate of generation of solute per unit volume of fluid in the reaction, n_W is the number of moles of water generated per mole of solute generated and V is the water molar volume. The concentration of solute per unit volume of fluid produced in the reaction (in the case of the dolomite reaction (5.12a), Ca^{2+} in solution) is specified by (5.8), namely,

$$\frac{\partial c}{\partial t} + \nabla \cdot (\bar{\mathbf{v}} c) - D \nabla^2 c = Q_C, \tag{5.45}$$

where the dispersion term is retained in anticipation that the solute concentration may vary abruptly inside the reaction zone. If one mole of incoming solute is removed from solution per mole of solute produced, an identical equation with the sign of Q_C reversed applies to that solute. Finally, the rate of change of molar concentration (per unit volume of the matrix) of the immobile solid material consumed by the reaction is specified by (5.9):

$$\frac{\partial s}{\partial t} = -\phi n_s Q_C, \tag{5.46}$$

where n_s is the moles of solid consumed per mole of solute produced in the reaction and ϕ is the porosity, the fraction of the matrix volume occupied by reactant. The porosity may change during the course of the reaction, a process considered later.

5.4.1 The front propagation speed and the fluid–rock ratio

Assume that the mean interstitial flow is locally unidirectional, with the local mean streamline taken to define the x direction. In the present context, the rate of generation of liquid water is extremely small compared with the hydrological flux divergences, so that $\bar{v}$ can be taken constant. The concentration pattern of the ions entering the region and undergoing reaction is specified by

$$\frac{\partial c}{\partial t} + \bar{v} \frac{\partial c}{\partial x} - D \frac{\partial^2 c}{\partial x^2} = Q_C \tag{5.47}$$

and (5.46) remains the same. Since Q_C is a function of c and s at given pressure and temperature, these equations permit solutions of the form $c = c(\xi)$, $S = s(\xi)$, where $\xi = x - Ut$, representing distributions that move through the matrix in the direction of the streamlines, with a speed U that is to be determined. Since, from the chain rule of differentiation, $\partial/\partial t = \partial/\partial\xi(\partial\xi/\partial t) = -U\partial/\partial\xi$ and $\partial/\partial x = \partial/\partial\xi$, equations (5.46) and (5.47) become

$$-Uc' + \bar{v}c' - Dc'' = Q_C,$$
$$Us' = \phi n_s Q_C, \tag{5.48}$$

the primes denoting differentiation with respect to ξ. Since the source Q_C of solute involved in the reaction (Mg^{2+} in the dolomite reaction) is a function of the local molar concentrations, c of dissolved solute and s of solid reactant, these are coupled differential equations that specify the spatial distributions of c, and s, as well as the speed U of propagation of the pattern through the matrix.

The distributions of solute concentration and mineral formation through the reaction zone certainly involve the chemical kinetics, represented by Q_C in the equations (5.48), but integral or overall properties such as the speed of propagation of the reaction front and the fluid-rock ratio (the volume of solute needed to produce unit volume of altered rock) depend only on the *total changes that occur across the reaction front* and are independent of the kinetics. To show this, the balance equations (5.48) are integrated along a streamline through the reaction front to obtain statements about the total changes or "jumps" in fluid volume flux, solute concentration and molar abundance of solid minerals involved. The suffix 0 is used to denote quantities upstream of the reaction zone and E is used to denote those downstream where the fluid has returned to its new equilibrium with the altered matrix.

As illustrated in the flux diagram in Figure 5.7, far upstream of the reaction zone, the concentration c of dissolved solute is c_0, the unaltered concentration of the source of fluid (i.e. the magnesium ions in the dolomite reaction). Moreover, upstream of the reaction zone, the mineral involved in the reaction (calcite) has already been replaced (by dolomite), so that $c \to c_0$, $s \to 0$, and $\bar{v} \to \bar{v}_0$ as x (or ξ) $\to -\infty$. Ahead, or downstream of the zone, the fluid has reached its new equilibrium with the unconsumed mineral, so that $c \to c_E$, say, while $s \to s_0$, the initial molar concentration (of calcite), and since the rate of chemical generation of water is small compared with the through-flow, $\bar{v} = \bar{v}_0$ throughout. In the reaction front, Mg^{2+} is disappearing from solution and if this goes to completion $c_E = 0$, but in any event, an integration of the solute balance with these boundary conditions gives

$$(\bar{v}_0 - U)(c_0 - c_E) = [Q_C], \tag{5.49}$$

which equates the net flux the of the active ions into the front to the rate at which they are reacting (to produce dolomite). Note that on either side of the reaction

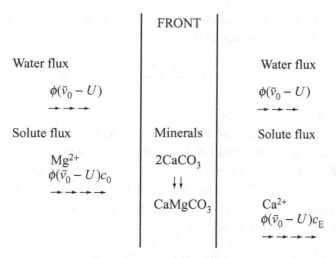

Figure 5.7. Mineral and solute balances across a reaction front, in a frame of reference moving to the right with propagation speed U through the matrix. In this frame, the flow and fluxes are steady in time. Horizontal arrows indicate physical fluxes and vertical arrows indicate chemical reactions.

zone, the fluid concentration gradients along the streamlines are zero, so that the integral of the second derivative, diffusion term vanishes. Within the reaction zone, reactant may be redistributed in space by dispersion but not generated nor destroyed by dispersion. Also, from the second of (5.48),

$$s_0 = (\phi n_S / U)[Q_C], \tag{5.50}$$

which specifies that the reaction is complete when all of the accessible original solid solute has been consumed in the reaction.

The kinetic term $[Q_C]$ can be eliminated between the two equations (5.49), and with (5.50), this yields a balance among the flux of solute into and out of the reaction zone and the rate at which the solid reactant is overtaken by it, as illustrated in Figure 5.7.

$$(\bar{v}_0 - U)(c_0 - c_E) - \frac{U}{\phi n_S} s_0 = 0. \tag{5.51}$$

Finally, this can be rearranged to give for the ratio of the front speed to the mean interstitial fluid velocity behind the front

$$\frac{U}{\bar{v}_0} = \frac{c_0 - c_E}{(c_0 - c_E) + (s_0/\phi n_S)} \approx \frac{\phi n_S(c_0 - c_E)}{s_0}, \tag{5.52}$$

since c_0, $c_E \ll s_0$ in the denominator of the middle term. The ratio of front propagation speed to the transport velocity u is

$$\frac{U}{u} = \frac{n_S(c_0 - c_E)}{s_0}, \tag{5.53}$$

which is very small.

The fluid–rock ratio r in these reactions is defined as the volume of fluid with which a unit volume of rock has reacted during the passage of the reaction front. Relative to the reaction front the interstitial fluid moves with speed $(\bar{v}_0 - U)$ but only over the fraction ϕ of the cross-sectional area, so that the fluid flux per unit area into the front is $\phi(\bar{v}_0 - U)$. The rate at which rock volume is overtaken by the front per unit area is simply U, so that the fluid–rock ratio is

$$r = \frac{\phi(\bar{v}_0 - U)}{U} = \frac{u_0}{U} - \phi, \tag{5.54}$$

where u_0 is the Darcy transport velocity. This is numerically very large. Note that some authors, unmindful of the fact that u_0 does not transform to a moving frame in the same manner as a true velocity vector does (see Section 2.3), give 1 rather than ϕ as the last term, but since $u_0/U \gg 1$, the numerical error is small. From (5.54), the fluid–rock ratio can be expressed to sufficient accuracy as

$$r \approx \frac{s_0}{n_S(c_E - c_0)},$$

$$= \left(\frac{\phi[Q]}{U}\right)\frac{1}{c_E - c_0},$$

from (5.46). Now $\phi[Q]\,dt$ represents the total production of volatile solute in the zone in the time interval dt and the volume of reaction is $U\,dt$ so that $\phi[Q]/U$ represents the moles n_V of volatile species released per unit volume of rock during metamorphism and the fluid-rock ratio is given alternatively by

$$r = \frac{n_V}{c_E - c_0}, \tag{5.55}$$

which is operationally equivalent to the expression given by Ferry (1987). If the generation of water in the reaction is regarded as significant, an additional factor $(1 - n_W V c_0)$ is included in the numerator, where n_W is the number of moles of water generated in the reaction per unit volume of fluid, V represents the molar volume of water and c_0 and c_E are the initial and final equilibrium molar concentration of the reacting solute far upstream and far downstream of the front. In this context these are generally numerically very small while n_V is of order unity.

5.4.2 *Profiles in the reaction front*

Although, as we have seen, many important aspects of the reaction front scenario are independent of the detailed reaction kinetics, the initial formation and propagation can be illustrated by a numerical calculation in which the reaction kinetics are (necessarily) prescribed. Suppose for the purpose of this illustration (i) that no water is *produced* by the reaction, $n_W = 0$, so that $\bar{v} = $ const., and (ii) that the source of solute per unit volume of fluid is a bilinear function of the relative concentration of fluid reactant generated in the front $X = c/c_E$ and of $S_N = s/s_0$, the number of moles s per unit volume of the reacting solid at any point as a fraction of the number initially present, s_0. Thus,

$$Q_C = -\gamma n_S c_E (1 - X) S_N \tag{5.56}$$

where $S_N - s/s_0$. Equations (5.47) and (5.46) then become

$$\frac{\partial X}{\partial t} + \frac{\partial}{\partial x}(\bar{v}X) - D\frac{\partial^2 X}{\partial x^2} = \gamma(1 - X)S_N \tag{5.57}$$

and

$$\frac{\partial S}{\partial t} = -(\phi n_S \gamma c_E / s_0)(1 - X)S_N. \tag{5.58}$$

Suppose that the fluid starts moving through the matrix at time $t = 0$, so that initially $S_N = 1$ throughout. The incoming fluid has $X = 0$ at $x = 0$ for all times t, and far from the interface the fluid is in equilibrium with the matrix, so that $c = c_E$ and $X = 1$.

The simultaneous solution of the coupled nonlinear equations (5.57) and (5.58) specifies the evolution of the distribution $s(x, t) = s_0 S_N(x, t)$ of the solid reactant and that of the relative concentration $X(x, t)$ of the interstitial fluid. The equations can be rewritten in a form suitable for computation by taking as a time scale, the depletion time (5.43), i.e. $T_D = (n_S \phi \gamma)^{-1}(s_0/c_E)$, and as a length scale the equilibration length $l_E = \bar{v}/\gamma$. Note, incidentally that the velocity scale for the front

$$\frac{l_E}{T_D} = \bar{v} n_S \phi(c_E/s_0) = U, \tag{5.59}$$

in accordance with (5.52) with $c_E = 0$. With $\tau = t/T_D$ and $\xi = x/l_E$, these equations reduce to

$$\left(\frac{U}{\bar{v}}\right)\frac{\partial X}{\partial \tau} + \frac{\partial X}{\partial \xi} - \left(\frac{D}{\bar{v}l_E}\right)\frac{\partial^2 X}{\partial \xi^2} = (1 - X)S_N \tag{5.60}$$

and

$$\frac{\partial S_N}{\partial \tau} = -(1 - X)S_N. \tag{5.61}$$

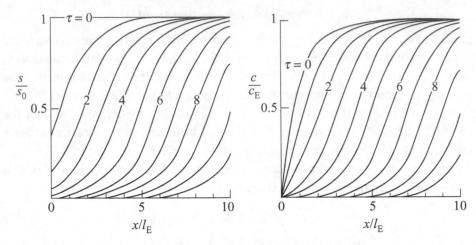

Figure 5.8. The development of an isothermal reaction front, with successive distributions of solid reactant is shown on the left and interstitial fluid concentration, on the right. The distance scale is the equilibration length and the time scale is the depletion time (5.43). Initially, at time $\tau = 0$, the solid mineral is uniform in the matrix and the interstitial fluid approaches equilibrium with the mineral over the equilibration or reaction length, as in Figure 5.5. Profiles are shown for equal time intervals $T_D = s_0/n_S \phi \gamma c_E$; by the time $\tau = 2$, the solid mineral is becoming seriously depleted near the interface and by $\tau = 4$, the depletion is vertically complete. The profiles of c/c_E and s/s_0 have become essentially identical, and they subsequently move away to the right without further change in shape, as the reaction front propagates through the fabric.

In this pair of dimensionless equations, the terms without coefficients are numerically of order unity, but the bracketed coefficients are small. From (5.56), it is evident that the front propagation speed U is very much less than the interstitial fluid speed $\bar{v}$, essentially because the solute concentrations are very much less than the abundance of the dissolving mineral. Also, in a "sandbank" medium, the dispersion coefficient is $\bar{v} \alpha_D$, where in a natural medium α_D is the autocorrelation length scale of the random variations in local permeability of the medium, the "dispersivity." Measurements in sandy aquifers on Cape Cod and in Denmark (Hess *et al.*, 1992, Jensen *et al.*, 1993) described in Section 3.3 gave values generally of the order of 0.01–1 m, considerably smaller than the equilibration length l_E so that $D/\bar{v} l_E = \alpha_D/l_E \ll 1$, though it is not known how representative these measurements are. Nevertheless, the second derivative term represents the smoothing effect of dispersion on the internal profiles and can probably be neglected without overall loss.

The evolving patterns, calculated from (5.60) and (5.61) are shown in Figure 5.8. Incoming fluid crosses the interface at $x = 0$ from the left, and the curves represent profiles of $S_N = s/s_0$ and $X = c/c_E$ after equal time intervals T_D. Initially, at times very short compared with the depletion time, the upper line and the curve marked 0, the entering interstitial fluid concentration quickly approaches its equilibrium value within a distance of the order of the equilibration length l_E as described in Section 5.3, but the abundance of the dissolving mineral is essentially unchanged. After a time $t \sim T_D$, ($\tau \sim 1$), however, the solid mineral is becoming depleted near the interface and the interstitial fluid concentration profile is flattening and beginning to move inward. By the time $t / T_D \sim 3$ or 4, the solid reactant is essentially depleted near the interface and both profiles have assumed an elongated S-shape which then moves through the matrix without further change in shape. The steadily propagating profiles of s/s_0 and c/c_0 can, in fact, be found analytically with the terms in parenthesis neglected. It is found that

$$\frac{c}{c_0} = \frac{s}{s_0} = \left\{ 1 - \exp \left[\frac{-\gamma(x - Ut)}{\bar{v}} \right] \right\}^{-1} . \tag{5.62}$$

These profiles are indistinguishable from those of Figure 5.8 when the dimensionless time τ is greater than about 4.

In these calculations, the particular form of these profiles once they have separated from the boundary (but not their propagation) involves the assumption that the dimensionless reaction rate is of the form $(1 - X)S_N$. Other assumptions will modify the profile somewhat, but more far reaching variations are produced by different flow patterns and in fracture–matrix media.

5.4.3 Reaction fronts in fracture–matrix media

One of the basic fluid scenarios involved in the formation of dolomite in the Triassic Latemar massif in Northern Italy is believed to involve the propagation of reaction fronts through an extensively fractured medium, a belief that is reinforced by a comparison between the expected characteristics of reaction fronts of this kind and the patterns of rock alteration observed and measured in the field.

The analysis combines the balances used earlier in simple reaction fronts with the flow characteristics of fracture–matrix media. Suppose that magnesium-rich sea-water enters an extensively fractured limestone bed and develops a reaction front as described earlier. As again illustrated in Figure 5.7, the sea-water containing c_0 moles per unit volume of magnesium ions Mg^{2+} moves at the interstitial, fracture network velocity $\bar{v}_F$ and passes through the newly created dolomite behind the reaction front before catching up with the front itself. As it enters the front region, the fluid encounters unaltered limestone along the fracture walls, where it loses Mg^{2+} and acquires Ca^{2+} as the limestone is replaced by dolomite in the reaction (5.12a). Emerging from the front, the fluid is now depleted of magnesium but is calcium-rich, and so passes through the limestone ahead of the front without further reaction. Fluid moves much more slowly into and through the lower porosity, lower permeability matrix blocks. The more rapid fracture flow exchanges fluid and solutes with the blocks, which contain most of the fluid. There, it is moving much more slowly so that the reaction can proceed further toward equilibrium before the solution is discharged into the next fracture downstream.

There is no net water generated or absorbed in this reaction, so that the mean velocity through the system is uniform and prescribed by external dynamics. There are three balances that do need to be considered – the flow of magnesium ions in the fracture system and in the blocks (and the exchanges between the two) and the rate of deposition of dolomite. The patterns of calcium ion flow and limestone depletion are the mirror images of the other three, and do not need to be considered separately.

The physical fluxes involved in the reaction (5.12a) above are also illustrated in Figure 5.7. The magnesium ion balance in a matrix block expresses, as in (3.108), the rate of change of ion concentration (moles per unit volume of the medium) in the block as the sum of the net exchange of solute with the surrounding fracture network – influx at concentration c_F and efflux at concentration c_M – together with the depletion of magnesium ions inside the block as a result of the reaction summarized above:

$$\phi_M \frac{\partial c_M}{\partial t} = \phi_M \frac{\overline{v}_M}{l_B}(c_F - c_M) - \phi_M Q, \tag{5.63}$$

where $\overline{v}_M / l_B = E$ is the exchange rate, the inverse of the mean time taken for fluid to traverse a block of mean dimension l_B. The source term Q is given by the kinetics as a function such as (5.60) of the relative concentrations, the density of reaction sites and proportions of reacting minerals, temperature, etc.

The magnesium ion balance in the connected fracture network, (3.109), also per unit volume of the medium, includes (a) advection through the network by the interstitial fluid (b) the net gain of solute from the blocks, which is equal in magnitude but opposite in sign to the exchange in the previous equation, as well as (c) ion depletion by reaction in the fractures

$$\phi_F \frac{\partial c_F}{\partial t} + \phi_F \overline{v}_F \frac{\partial c_F}{\partial x} = -\phi_M \frac{\overline{v}_M}{l_B}(c_F - c_M) - \phi_F Q. \tag{5.64}$$

The first term involving ϕ_F ($\sim 10^{-4}$) is very small and often omitted (Barenblatt *et al.*, 1990). Note particularly that the porosity ϕ_M and transport velocity $\phi_M \overline{v}_M$ of the *matrix blocks* is involved in the exchange term. The sum of the last two equations, in the form

$$\frac{\partial(\phi_M c_M + \phi_F c_F)}{\partial t} = -(\phi_M + \phi_F)Q - \phi_F \overline{v}_F \frac{\partial c_F}{\partial x}, \tag{5.65}$$

expresses the rate of change of total Mg^{2+} ions in any unit volume of fabric as the rate of loss of ions from the fluid in the dolomite reaction and the net efflux rate from the volume by advection in the fractures. The exchange terms balance out, as they must.

Finally, the rate of increase of the number of moles of solid dolomite $CaMg(CO_3)_2$ formed per unit volume is equal to the number of moles of magnesium ions per unit volume disappearing from the solution, as given in the last equation:

$$\frac{\partial s}{\partial t} = (\phi_M + \phi_F)Q. \tag{5.66}$$

The propagation speed of the front can be found in essentially the same way as for a classical "sandbank" medium. The fracture storage component in the first term of (5.64) can be neglected, and with the insertion of (5.66), it becomes

$$\frac{\partial}{\partial t}(\phi_M c_M + s) = -\phi_F \bar{v}_F \frac{\partial c_F}{\partial x}, \tag{5.67}$$

which is a conservation statement that the rate of increase at any point of the number of moles containing magnesium is equal to the negative gradient of the flux of magnesium ions in the fractures per unit volume. If the distributions are all functions of $\xi = x - Ut$, this equation can be integrated as before to give

$$-U(\phi_M c_M + s) + \bar{v}_F \phi_F c_F = \text{const.}$$

Downstream of the front (ahead of it), there is no magnesium from the seawater and no dolomite, so that $c_M = c_F = s = 0$, and the constant is zero. Upstream, the entering seawater has not yet reacted and $c_M = c_F = c_0$, its initial molar concentration of Mg^{2+} ions, while $s = s_0$, the number of moles per unit volume of dolomite produced in the passage of the front. (This may be somewhat less than the number of moles of calcite originally present, since some may have been embedded deeply inside blocks and were inaccessible to the incoming solution.) Thus the propagation speed of the front is

$$\frac{U}{v_F} = \frac{\phi_F c_0}{\phi_M c_0 + s_0} \approx \frac{\phi_F c_0}{s_0}, \tag{5.68}$$

where $c_0 \ll s_0$. This equation is structurally similar to, but different in detail from the corresponding result (5.52) and (5.53) for a "sandbank" medium. In this case, no water is produced in the reaction and $n_W = 0$, while the number of moles of dolomite produced per mole of Mg^{2+}, $n_S = 1$. In both cases, the numerator specifies the change in solute moles being transported to and from the front; in the fracture–matrix flow, this occurs almost entirely in the fractures.

The thickness l_E^* of the front, the fracture–matrix equilibration length, can also be estimated simply. In the overall solute balance (5.65) applied to the region of the front, the time derivative is of order U/l_E^*, so that the first term is of order $U\phi_M c_0/l_E^*$, while the flux gradient represented by the last term is of order $\phi_F \bar{v}_F c_0/l_E^*$, much larger in view of the smallness of the ratio (5.68). The balance in (5.65) must then be essentially between the sink of Mg^{2+} ions, proportional to $\phi_M \gamma c_0$ and the solute flux gradient, $\phi_F \bar{v}_F c_0/l_E^*$. On equating these, we obtain

$$l_E^* = \frac{\phi_F \, \bar{v}_F}{\phi_M \, \gamma} = \frac{u_F}{\phi_M \gamma}. \tag{5.69}$$

Implicit in this result is the assumption that the kinetic reaction rate γ is larger than the exchange rate between blocks and fractures E, so that when fluid enters a block, the reaction has time to approach completion before rejoining the fracture flow beyond. If, on the contrary, $E > \gamma$, similar arguments applied to (5.64) lead to

$$l_E^* = \frac{u_F}{\phi_M E}.$$

Compare these with the previous expression (5.26) for the equilibration length in a "sandbank" medium, namely

$$l_E = \frac{\bar{v}}{\gamma} = \frac{u}{\phi \gamma},$$

where u is the Darcy transport speed. In the fracture–matrix medium, the transport occurs in the fractures and the storage in the matrix blocks, so that it is no surprise that the fracture–matrix equilibration length scale depends on the transport speed in the fractures and the porosity of the matrix blocks, as (5.69) shows. Note also that when $n_S = 1$ (one mole of mineral dissolved per mole of solute produced) the front thickness scale (5.69) divided by the front speed $U = u_F c_0/s_0$, from (5.68), is just the depletion time (5.43), namely,

$$T_E = \frac{s_0}{\phi_M \gamma c_0}, \tag{5.70}$$

again involving the matrix porosity.

5.4.4 Sorbing contaminant plumes

Reactive groundwater contaminants such as ammonium from wastewater and nitrogen or phosphorus from excess fertilization are retarded by their attachment and interaction with the solid matrix and move more slowly then the interstitial fluid velocity. The retardation factor, usually defined (in the inverse form) as the ratio of the mean interstitial fluid velocity to the speed of advance of a contaminant front, is therefore larger than one. The retardation effect has been demonstrated in field measurements by Ceazan, Thurman and Smith (1989) and measured in greater detail by Böhlke, Smith and Miller (2006) in work described below. The simplest of the physical-chemical processes involved is sorption, in which dissolved ions are temporarily bound to the matrix when the solution concentration is large, and flushed from it when small. Microbially induced transformations that depend upon the aquifer geochemistry and configuration may have similar consequences. The "retardation factor" is essentially equivalent to the fluid–rock ratio (5.58) in geomorphology, which specifies the volume of fluid with which unit volume of rock has reacted since the passage of a reaction front, and is generally of order 10^4 or so. In groundwater contamination on the other hand, involving much more soluble phosphate and ammonium ions, the immobile ions in the matrix are derived from the interstitial solution by sorption, the contaminant mass per unit volume in the liquid and solid phases are comparable and the retardation factors are usually in the range 2–10.

The physical-chemical balances involved in this context have the same general form as (5.44), (5.45) and (5.46), but with simplifications. No water is generated geochemically, so that $n_W = 0$ and the continuity equation is $\nabla \cdot \bar{\mathbf{v}} = 0$. The total mass c_L of solute in the liquid per unit volume of the fabric (solid plus pores) is ϕc_L, so that the solute balance is

$$\phi \left\{ \frac{\partial c_L}{\partial t} + \bar{\mathbf{v}} \cdot \nabla c_L - D \nabla^2 c_L \right\} = -Q, \tag{5.71}$$

where the macroscopic dispersion coefficient $D = \bar{v} \alpha_D$ is the product of the mean interstitial fluid velocity and the dispersivity α_D, the characteristic scale of medium variability (as in Sections 2.10 and 3.3), which in a sandy aquifer is usually of order a meter or less (Hess *et al.*, 1992). The source term Q is the rate of sorption of ions from solution to the solid matrix (mass of solute per unit volume of matrix per unit time).

The absorbed solute balance for the fraction $(1 - \phi)$ of the fabric volume, is

$$(1 - \phi) \frac{\partial c_S}{\partial t} = Q \tag{5.72}$$

where c_S is the mass per unit volume of absorbed solute in the matrix. The sum of these does not, of course, involve the exchange rate Q. If the absorbed solute does not penetrate throughout the individual grains, the coefficient in this last equation should properly be somewhat smaller than $(1 - \phi)$.

The exchange rate Q from fluid to solid vanishes when $c_L = c_S$, is positive when the interstitial fluid is more concentrated, $c_L > c_S$, and is negative when $c_L < c_S$, suggesting that, near equilibrium, $Q \approx \gamma(c_L - c_S)$, where γ is here the reaction rate for sorbtion. Thus

$$\phi \left\{ \frac{\partial c_L}{\partial t} + \overline{\mathbf{v}} \cdot \nabla c_L - D \nabla^2 c_L \right\} = -\gamma(c_L - c_S) = -(1 - \phi) \frac{\partial c_S}{\partial t} \qquad (5.73)$$

which are respectively of order

$$c_L \overline{v}/l \qquad c_F \alpha_D \overline{v}/l^2 \qquad \gamma \Delta c \qquad c_S \overline{v}/l.$$

The salient characteristics of the reacting plume or contaminated region can be found without detailed calculation from the structure of these balances. In a sandy medium, the porosity is about 0.4, which is of order unity, and inside the curly brackets, the first two (advective) terms are of order $c_L \overline{v}/l$, where l is the characteristic dimension in the flow direction of the contaminant cloud. The macroscopic dispersion term in a typical aquifer is of relative order $\alpha_D/l \sim 10^{-3}$. Although the characteristic aquifer-scale horizontal flow divergence and spreading (Section 3.3) extends a contaminant cloud primarily in the flow direction, Fickian dispersion is usually negligible in this balance and will henceforth be neglected. Relative to the first terms, the central transfer term is of order

$$\frac{\gamma l}{\overline{v}} \frac{\Delta c}{c_L} = \frac{l}{l_E} \frac{\Delta c}{c_L}, \qquad (5.74)$$

where l_E is the equilibration length (5.20) and Δc represents the (as yet unknown) general magnitude of the difference in concentration $(c_L - c_S)$ between interstitial fluid and solid matrix.

Consider the anatomy of these balances. When $l/l_E = \gamma l/\overline{v} \ll 1$, the contaminant cloud is so small compared with the equilibration length and the sorption rate so slow that contaminant in the interstitial fluid scarcely affects the matrix even if the concentration difference Δc is comparable with c_D – the interstitial fluid and the matrix are chemically uncoupled, as discussed in Section 3.3.

The transfer term alone cannot be an order of magnitude larger than the other terms if the equation is to balance, so that in all interesting cases,

$$\frac{\gamma l}{\overline{v}} \frac{\Delta c}{c_L} = 0(1) \quad \text{so that} \quad \frac{\Delta c}{c_L} \sim \frac{\overline{v}}{\gamma l} = \frac{l_E}{l}. \qquad (5.75)$$

The quantity $\gamma l/\overline{v}$ also expresses the ratio of the reaction rate γ to the rate $\overline{v}/l$ at which interstitial fluid moves through the contaminant patch of dimension l in the flow direction. If $\gamma \gg \overline{v}/l$, the reaction is fast to respond to changing conditions. The equilibration length $l_E = \overline{v}/\gamma$ as in (5.20), and when this is small compared with the dimension of the contaminant cloud in the direction of flow,

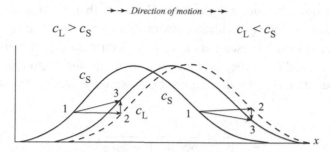

Figure 5.9. The mechanism of retardation produced by sorption and desorption. An injected pulse of interstitial fluid is advected at the interstitial fluid velocity to the dashed profile, labeled 2. However, the solute concentration at this point has been decreasing and solute has been partly absorbed by the matrix (3) producing a phase lag so that the profile advances more slowly.

the relation (5.75) asserts that the matrix and the interstitial fluid are *close to equilibrium* everywhere, with only small concentration differences (mass per unit volume) between interstitial fluid solid matrix, $\Delta c = |c_L - c_S| \ll c_D$. The cloud moves more slowly than the interstitial fluid, and at the rear of the patch, the concentration of the entering interstitial fluid is somewhat less than the value for local equilibrium with what has already been sorbed into the matrix, $c_L < c_S$, as illustrated in Figure 5.9. Solute is desorbed from the matrix to the fluid and its concentration increases as it moves forward. At the centre of the patch the two are in equilibrium and in the leading sections $c_L > c_S$ so that contaminant is sorbed from the fluid to the surrounding matrix and the distribution of c_S also moves ahead.

These considerations can be quantified readily. In a contaminant release over a finite time interval T_R, a cloud of length $l \approx \bar{v} T_R$ is produced; the length increases as the flow moves toward discharge. When the equilibration length is small (say 10^{-2}) compared with the patch length, the concentration differences are also small, $\Delta c = |c_L - c_S| \ll c_F$, and

$$\phi \left\{ \frac{\partial c_L}{\partial t} + \bar{v} \frac{\partial c_L}{\partial x} \right\} = (1 - \phi) \frac{\partial c_S}{\partial t} = (1 - \phi) \frac{\partial c_L}{\partial t} (1 + O(l_E/l)), \quad (5.76)$$

where the x-direction is that of the flow and the interstitial velocity $\bar{v}$ can be considered to be constant over distances small compared with the aquifer length. The first and last groups involving first derivative space and time terms constitute a wave equation in the fluid concentration c_L, with the solution

$$c_L = \hat{c}_L f(x - U_C t), \quad c_S = \hat{c}_S g(x - U_C t), \quad (5.77)$$

where the contamination profiles f and g are determined not by the equation (5.76) but by the spatial distribution and the time history of the contaminant release. If it occurred at a single site over a limited time duration, then f and g are non-zero for a limited space-time interval with maximum values that can be taken as one. The expression (5.77) then represents a cloud with maximum concentration $\hat{c}_L$ in the interstitial liquid and $\hat{c}_S$ in the solid matrix, moving in the x-direction with speed U_C. This propagation speed is determined by substitution into (5.76), whence, after cancellation,

$$\frac{U_C}{\bar{v}} = \frac{\phi c_L}{\phi c_L + (1-\phi)c_S}, \qquad (5.78)$$

In words,

$$\frac{wave\ speed}{fluid\ speed} = \frac{mass\ of\ dissolved\ (mobile)\ solute\ per\ unit\ volume\ of\ fabric}{total\ mass\ of\ solute\ per\ unit\ volume\ of\ fabric}.$$

The propagation speed of the contaminant cloud is clearly less than the mean interstitial fluid velocity. The "retardation factor" is defined as the inverse of this:

$$R = \frac{\bar{v}}{U_C} = 1 + \frac{(1-\phi)c_S}{\phi c_L} = 1 + K_d, \qquad (5.79)$$

where

$$K_d = \frac{mass\ of\ sorbed\ solute\ p.\ u.\ vol}{mass\ of\ mobile\ solute\ p.\ u.\ vol}.$$

K_d is called the "volumetric distribution coefficient." These expressions are given by Freeze and Cherry (1979) and by Böhlke *et al.* (2006) in slightly different notation. They can be obtained alternatively by considering the mass balances across a moving contaminant front.

Note that when the sorption reaction time γ^{-1} is small compared with the time $l/\bar{v}$ for interstitial fluid to pass through the length l of the cloud of contaminant (in the field, a decade, possibly), the concentrations (mass per unit volume) c_L in the fluid and c_S in the solid matrix are almost equal and close to mutual equilibrium:

$$c_S = c_L \{1 + O(l_E/l)\}$$

(where $l_E/l \ll 1$). In this limit, there is very little *net* interchange of solute between the fluid and solid. When $c_L = c_S$, the retardation factor (5.79) reduces to

$$R \approx 1/\phi, \qquad (5.80)$$

and $U_C \approx \phi\bar{v}$, the propagation speed of the cloud is approximately equal to the usual transport velocity within the same fractional error bound.

Ceazan *et al.* (1989) and Böhlke *et al.* (2006) describe valuable and informative sets of measurements on ammonium transport and reaction in the contaminated groundwater plume created by the injection of treated wastewater into the aquifer

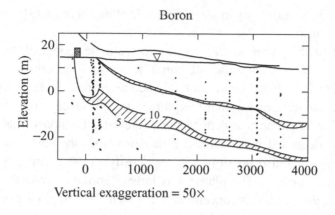

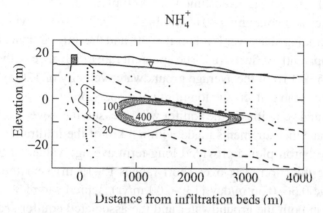

Figure 5.10. Vertical longitudinal sections through the wastewater plume from the Massachusetts Military Reservation in 1994, showing the concentrations (μmol/l) of boron in the upper panel and ammonium ion in the lower one. The boron, which is not reactive, delineates the plume boundary, while the ammonium cloud moves more slowly because of sorption with the solid matrix. The top curves represent the approximate land surface above the water table, which is indicated by the triangle.

beneath the Massachusetts Military Reservation in western Cape Cod over the period 1936 to 1995. The aquifer is largely glacial outwash, with quartz and feldspar grains and having a porosity of about 0.4. The crest of the water table in this region is approximately 6 km NNE of the decommissioned sewage infiltration beds shown in Figure 5.10 (beyond the left of the region shown). The mean groundwater flow is generally radial from this crest and moves through the measurement site in a south-south-westerly direction. Because of rainwater infiltration the mean flow speed might be expected to increase very roughly linearly with distance from the groundwater crest, although there are considerable variations because of flow fingering from inhomogeneity in the medium (LeBlanc *et al.*, 1991) and local flow deflection by ponds.

The extent of the plume in 1994 shown in the upper panel of this figure was delineated by high concentrations of boron, a common constituent of domestic wastewater that is both stable and mobile and therefore a useful tracer. The vertical longitudinal sections through the plume are reproduced from Böhlke *et al.* (2006). The boron plume that defined the wastewater boundary was continuous, and gradually deepened with increasing distance from the source as a result of aquifer recharge and also possibly a water density in the plume slightly greater than that in the ambient. The plume was largely anoxic. Inside the wastewater pulse or cloud, ammonium is concentrated in a well-defined cloud that has left the infiltration beds but has lagged behind the boron plume front. The rate of disposal of treated wastewater containing NH_4^+ was highest ($c. 2 \times 10^6$ m^3/yr) between 1941 and 1945, continuing at a lower level until 1970, and the longitudinal distribution of ammonium apparently reflects this. The apparent rate of advance of the ammonium cloud was 0.25 ± 0.10 of the average groundwater velocity of 120 m/yr. The large range about the mean reflects the longitudinal dispersion.

Interstitial fluid velocities in the aquifer were measured in several different ways. Groundwater age measurements made within 3 km of the infiltration beds, in and adjacent to the boron plume provided long-term average values of 90–120 m/yr, while bromide injection tests over short ranges (3–15 m) near the midpoint of the plume gave 0.36–0.56 m/day (130–200 m/yr). Paired samples of ammonium concentration in both the groundwater and the associated aquifer sediments were taken at several points in and adjacent to the cloud. Böhlke *et al.* reported their concentration data in terms of μmol per gram of solid or of water, whereas Ceazan *et al.* use μg N per gram of dry sediment for the sorbed ammonium and μg N per ml for the aqueous ammonium. When the results are expressed in consistent concentration density units, such as μmol/cc, the measured points from both sets of measurements group reasonably well, as shown in Figure 5.11, lying close to but slightly above the 45° line along which $c_F = c_S$. Although the amount of data is limited, this result appears to be consistent with the "fast reaction" limit $\gamma \gg \bar{v}/l$, with c_F slightly smaller than c_S. This may be characteristic of the trailing edge of the cloud where the solid is losing ammonium to the fluid.

Retardation factors inferred by Böhlke *et al.* from the migration of the concentration maximum and of the front scattered between 2.8 and 6.4, apparently reflecting medium inhomogeneity. Ceazan *et al.* estimated from the regional scale retardation of the ammonium front that $R \approx 2$, while from batch concentration measurements and the use of (5.79) they obtained $R \approx 2.5$. These are generally consistent with, but on average larger than the value 2.5 that would be expected

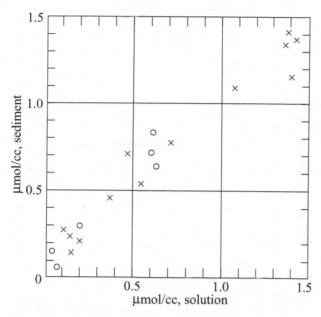

Figure 5.11. Paired measurements of NH_4^+ in groundwater samples (horizontal axis) and associated aquifer sediment extracts by Böhlke *et al.* (2006), "o", and Ceazan *et al.* (1989), "×".

from the "fast reaction" expression (5 80) with a porosity of 0.4. The reason for this difference is unclear, although it is possible that some biological processes may have contributed to the retardation in at least some cases.

5.5 The gradient reaction scenario

The reaction fronts described in the preceding section arise at mineralogical boundaries or interfaces and propagate in the flow direction; they separate an unaltered region ahead of the front from the region behind where the reaction has produced a different equilibrium between the altered host rock and the interstitial fluid. A second flow reaction scenario is found when the temperature and/or pressure vary along the streamlines of the flow domain, and even if the medium is chemically homogeneous, the equilibrium solute concentration varies as these ambient conditions change along the lines of flow. These variations promote flow-controlled reactions of another type, termed gradient reactions, that can occur *simultaneously* throughout the whole matrix or along systems of cracks at rates proportional to the flow speed and to the gradients of temperature and pressure in the direction of flow.

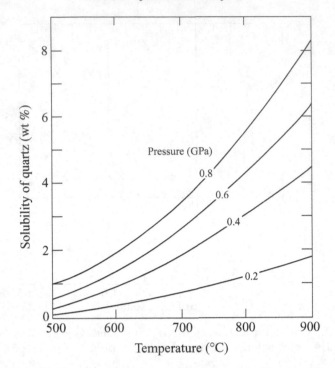

Figure 5.12. Solubility of quartz in supercritical water as a function of temperature and pressure, after Anderson and Burnham (1965) and Philpotts (1990). Note: a pressure of 1 GPa is equal to the weight of about 30 km of crustal rock.

If the pressure and the temperature vary along the flow path, the local equilibrium point also varies. As the fluid elements move through the medium, they interact with the matrix in response to the gradually varying conditions. Even though the composition of interstitial fluid may remain *the same in time* at each spatial point, it varies *along the flow trajectory* so that, as the fluid elements move along their paths, solute is continually being added to or taken from solution. As a consequence, the mineral composition is gradually altered along the flow path by reaction with the moving fluid, by dissolution of the matrix or mineralization of solute into it. In general, the equilibrium concentration of the interstitial fluid depends on the environmental variables of temperature, pressure and the dissolved concentrations $c_1, c_2, \ldots$ of other species (including hydrogen ions) that may influence the reaction. With few exceptions, the equilibrium concentration increases with temperature and pressure, $\partial c_E/\partial T > 0$ and $\partial c_E/\partial p > 0$ as shown in Figure 5.12. Below the surface of the earth the fluid temperature and pressure both generally decrease with vertical position, so that the vertical gradient of equilibrium concentration is generally negative, $\partial c_E/\partial z < 0$. In the mean, fluid elements move through the geochemical environment at the interstitial flow speed $\bar{v}$, and when the length scales of the environmental variations are large compared with the equilibration length $\bar{v}/\gamma$, one would expect intuitively that the actual chemical composition of the moving fluid elements would remain close to local equilibrium in their slowly varying environment. Wood and Hewett (1982) recognized this explicitly and assumed that $c \approx c_E$ to sufficient accuracy. This assumption is, in fact, faulty, but it has become known as "the equilibrium hypothesis" and, despite some conceptual difficulties (see Lichtner, 1991), it has been used extensively and successfully.

The solute balance equation (5.4), repeated here,

$$\frac{\partial c}{\partial t} + \bar{v} \cdot \nabla c - D \nabla^2 c = \gamma c_E f(r, X),$$

specifies the distribution of reacting solute in space and time in a given pattern of flow and it can be used to examine the accuracy of this hypothesis. To avoid extraneous complications, consider the simplest physical environment that combines the essential elements of an interstitial fluid undergoing reaction as it moves through a matrix of variable temperature and pressure. The pattern of solute concentration is assumed to be steady in time as it evolves in space, so that the first term is neglected. The solute dispersion effect is merely smooths out small-scale spatial solute irregularities, and we neglect this also. The solute balance is then between advection and the chemical kinetics of the solute–matrix interactions which are sensitive to spatial variations in temperature and pressure.

Accordingly, consider hot fluid *rising* steadily and, on average, vertically in the crust through fractures or through a permeable inclined layer, with the temperature and pressure both decreasing approximately linearly with height over intervals large compared with the equilibration length. Consequently, as in Figure 5.12, the *equilibrium* solute concentration also decreases in the vertical, $\partial c_E / \partial z < 0$. The solute balance above therefore reduces to one between vertical advection of solute and chemical deposition from solution,

$$\bar{v}\frac{\partial c}{\partial z} = Q = \gamma c_E f(c/c_E), \tag{5.81}$$

where Q represents the source, the rate of addition of solute (mass per unit volume of fluid per unit time) from the reaction, and $-Q$ represents the corresponding rate of deposition of solid mineral from the solution. From the kinetics of the reaction, the dimensionless function $f = 1$ when c/c_E is small (very dilute solutions), $f = 0$ when $c/c_E = 1$ (saturated solutions) and negative when $c/c_E > 1$ (supersaturated). We are most interested in conditions near local equilibrium, and a suitable form for f that has these properties is

$$f = 1 - c/c_E.$$

Thus, near equilibrium, (5.81) becomes

$$\bar{v}\frac{\partial c}{\partial z} = \gamma(c_E - c). \tag{5.82}$$

Let $\Delta c = c - c_E$ represent the difference between the solute concentration at any point and the equilibrium concentration for the temperature and pressure at that point, i.e. the "error" in the equilibrium hypothesis. Rewrite equation (5.82) in terms of (Δc):

$$\frac{\partial}{\partial z}(\Delta c) + l_E^{-1}(\Delta c) = \beta, \tag{5.83}$$

where $\beta = -\partial c_E/\partial z > 0$, the rate at which the equilibrium concentration decreases with vertical position because of decreasing temperature and pressure. The solution to (5.83), starting off the flow with complete equilibrium, $(\Delta c) = 0$, at an arbitrary level $z = 0$, is

$$(\Delta c) = \beta l_E \{1 - \exp(-z/l_E)\} > 0. \tag{5.84}$$

This was assigned the value zero at the initial level $z = 0$, but as the fluid moves upward, the fluid concentration relative to the local equilibrium increases towards βl_E as z increases beyond the equilibration length.

Thus, in upward fluid flow, even though the fluid is assumed to be in local equilibrium at the initial level, it becomes increasingly super-saturated as it rises and its temperature and pressure both decrease, as illustrated in the right-hand panel of Figure 5.13. The *equilibrium* concentration $c_E(z)$, indicated by the broken line, decreases with elevation and the actual concentration in the rising fluid (the continuous line) becomes super-saturated by the amount

$$c - c_E = \beta l_E = \left|\frac{\partial c_E}{\partial z}\right|\frac{\bar{v}}{\gamma}$$

leading to continuing mineral deposition in response to the vertical decrease in temperature and pressure as the solution moves through. After the fluid has risen a distance of the order of the equilibration length, the degree of super-saturation stabilizes at βl_E as the lines $c(z)$ and $c_E(z)$ become parallel. *The actual concentration of interstitial fluid and the equilibrium concentration are different (as is necessary for deposition), but after the initial adjustment, their vertical gradients become equal.* If the flow is downward as shown in the left-hand panel, the fluid is unsaturated by the same amount and the species concentration is augmented by dissolution from the walls.

The most important property of this solution is that on scales large compared with the equilibration length, the actual fluid concentration of interstitial fluids moving either upward or downward differ from the equilibrium concentration, but the

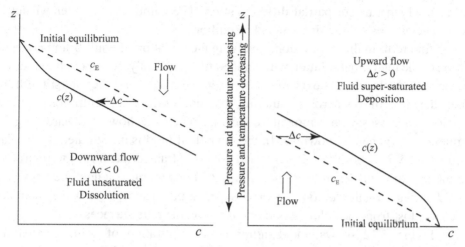

Figure 5.13. When interstitial fluid is at rest relative to the matrix containing a reactive solid, its concentration tends towards equilibrium at the local temperature and pressure. If the equilibrium concentration decreases with height, upwardly moving fluid becomes super-saturated and solute is deposited, while downwardly moving fluid becomes unsaturated and solid mineral dissolves from the matrix walls. The concentrations are offset from the equilibrium values, but the vertical *gradients* are the same.

vertical gradients of the two are the same. In (5.81), the rates of exchange between the solid matrix and the moving interstitial fluid involve the concentration *gradients* in the flow direction, rather than the concentrations themselves. Consequently, the "equilibrium hypothesis" that $c = c_E$, though erroneous, is not necessary and does not invalidate the Wood–Hewett statement that

$$Q = \overline{\boldsymbol{v}} \cdot \nabla c = \overline{\boldsymbol{v}} \cdot \nabla c_E, \tag{5.85}$$

where Q is the mass of solute added per unit volume of the fluid per unit time and the vertical gradients of c and c_E are equal.

5.5.1 Dissolution and deposition rates in gradient reactions

Re-written in terms of the transport velocity $\boldsymbol{u} = \phi\overline{\boldsymbol{v}}$, equation (5.85) now becomes a prescription for the source term Q – the rate of addition of species to the solution (moles per unit volume of the *fabric* per unit time). Correspondingly, the rate of mineral deposition or precipitation per unit volume *from* the solution is

$$\begin{aligned} Q_M &= -\boldsymbol{u} \cdot \nabla c_E(T, p, c_1, c_2, \ldots) \\ &= -\boldsymbol{u} \cdot \left\{ \frac{\partial c_E}{\partial T} \nabla T + \frac{\partial c_E}{\partial p} \nabla p + \sum_n \frac{\partial c_E}{\partial c_n} \nabla c_n \right\} \end{aligned} \tag{5.86}$$

from the chain rule for partial differentiation. The summation is over whatever other species $n = 1, 2, \ldots$ influence the equilibrium.

The first term in this expression, involving the variation of equilibrium concentration of the interstitial solution with temperature, is usually dominant. It expresses the rate at which solute is exchanged between the matrix and the interstitial fluid when the flow moves across isotherms. The equilibrium concentration of most solutions increases with temperature, so that when the flow is towards higher temperatures (generally downward), the interstitial fluid is unsaturated, the scalar product $\boldsymbol{u} \cdot \nabla T$ is positive, $Q_C > 0$ and solute is transferred from the matrix to the interstitial fluid. When the flow is down the temperature gradient, i.e., upward, $\boldsymbol{u} \cdot \nabla T$ is negative, the solution becomes supersaturated and $Q_C < 0$; the chemical species is lost from the solution and deposited on the pore surfaces.

As Figure 5.12 shows, the equilibrium concentration of solutes generally increases also with the fluid pressure. Even though the total fluid pressure may be close to lithostatic, the local *gradient* in total pressure in a connected region is primarily hydrostatic, so that $\boldsymbol{u} \cdot \nabla p = -\rho g u_Z$ where $\boldsymbol{u}$ is the vertical component of mean interstitial fluid velocity. The middle term on the right of equation (5.86) becomes

$$-\left(\frac{\partial c_E}{\partial p}\right) \rho g u_Z,$$

which is generally small compared with the first term.

The last term, involving the concentration gradients of other solutes, can be simplified considerably. If they do not enter into the reaction but may be purely inhibitory, their concentration remains the same along the flow path and this term vanishes. If any of them, say the nth, does enter, a separate species balance can be written for this species with a source term equal to $-Q_C$ times the moles consumed per mole of the solute in (5.86) generated, i.e.

$$\boldsymbol{u} \cdot \nabla c_n = -m_n Q_C.$$

The rate of mineralization (moles per unit volume per unit time) $Q_M = -Q_C$ becomes

$$Q_M = -\left\{1 + m_n \left(\frac{\partial c_E}{\partial c_n}\right)\right\}^{-1} \left\{\left(\frac{\partial c_E}{\partial T}\right) \boldsymbol{u} \cdot \nabla T - \left(\frac{\partial c_E}{\partial p}\right) \rho g u_Z\right\}. \quad (5.87)$$

At first sight it is perhaps surprising that the rate at which reaction or dissolution or precipitation occurs in this gradient reaction scenario is independent of the chemical kinetics, provided only that the equilibration length is small compared with the size of the region. The limiting factor in the overall process is the rate at which fluid moves through the system, providing fluxes of dissolved species into and out of each volume element at the matrix. Note that Q_C represents the rate of addition of the species to the solution (moles per unit volume per unit time) so that in dissolution, $Q_C > 0$, and in precipitation, $Q_C < 0$. In gradient reactions, mineralization is produced throughout a region both by interstitial flow in the direction of the local temperature gradient and by interstitial flow in the direction of the total pressure gradient, both of which are usually essentially vertical.

The applicability of the expression (5.87) is not limited to flow in permeable media and indeed the most visible evidence of mineral deposition by flow is the presence of networks of veins, largely of quartz, in many metamorphic rocks. The same formula can be applied to flow in a network of fractures or conduits, with u, u_Z representing the volume flux in the plane of the fracture or along the conduit and with Q_C and Q_M interpreted as the rate of exchange of solute with the matrix per unit area of the fracture or per unit length of the conduit, respectively. The result (5.87) can also be integrated over time and it retains the same form with Q_M^{tot} now representing the total density of mineralization (moles per unit volume) and V the total volume of fluid per unit cross sectional area that has moved through the region.

$$Q_M^{\text{tot}} = V \left\{ -\left(\frac{\partial T}{\partial z}\right)\left(\frac{\partial c_E}{\partial T}\right) + \rho g \left(\frac{\partial c_E}{\partial p}\right) \right\}, \tag{5.88}$$

where c_E is the molar concentration of the infiltrating fluid at equilibrium, a generally increasing function of temperature and pressure.

For example, quartz, SiO_2, is an extremely common vein material in metamorphic rocks. Although relatively insoluble in surface waters, its solubility increases significantly with increasing pressure and temperature, as shown in Figure 5.12. At a temperature of 650 °C and pressure of 0.375 GPa, the equilibrium concentration c_E is about 0.28 g-mol/l SiO_2, its rate of variation with temperature, $\partial c_E / \partial T \sim 1.6 \times 10^{-2}$ g-mol/l per °C and with pressure, $\partial c_E / \partial p \sim 0.83$ g-mol/l per GPa. Saturated fluid becomes supersaturated as it rises toward the surface because of the decreasing temperature and pressure, both of which reduce c_E. As a result, quartz is precipitated simultaneously along the length of the flow path, and in a previously fractured medium, a network of quartz veins is developed as described by Haszeldine, Samson and Cornford (1984) and McBride (1989).

In the calcite dissolution reaction

$$CaCO_3 + CO_2 + H_2O \Leftrightarrow Ca^{2+} + 2HCO_3$$

the rate of generation of Ca^{2+} ions per unit volume of interstitial fluid is equal to the rate of dissolution of calcium carbonate. Carbon dioxide in solution is consumed by the reaction and the term $\partial c_E / \partial c_n = (\partial c_n / \partial c_E)^{-1}$ represents the variations in equilibrium concentration of Ca^{2+} ions with concentration of dissolved

CO_2. The equilibrium concentration of Ca^{2+} is smaller when CO_2 is depleted, so that $\partial c_E / \partial c_n > 0$. Measured values for the variation of saturation solubility with temperature and total pressure are summarized by Barnes (1979). Calcite is an unusual solute whose equilibrium concentration *decreases* with increasing temperature. At a partial pressure of CO_2 of one atmosphere, at about 100 °C $(\partial c_E / \partial T) \sim -6 \times 10^{-6}$ g-mol/l per °C. The variation with pressure $(\partial c_E / \partial p)$ is positive, but its magnitude is insufficient to overcome the increase in solubility of a rising fluid moving down the geothermal temperature gradient. The sum of the two flow terms in (5.87) is therefore positive, and in upward flow, calcite dissolution will tend to occur.

5.5.2 The rock alteration index

In many reactions, the pressure term in (5.87) appears to be subordinate to the temperature term, and the *pattern* of mineral alteration within a given structure is determined primarily by the distribution of $u \cdot \nabla T$ associated with the flow. As suggested by Dr. James Wood (1987a,b), this quantity can conveniently be called the *rock alteration index*, or RAI. Distributions of this index are illustrated in Figures 4.11 to 4.14 and elsewhere in this book for various flow patterns. Regions where it is relatively high indicate sites of more rapid alteration by gradient reactions and, over time, sites of denser mineral accumulation or depletion.

Fault zones, particularly dilatational fault jogs connecting offset planes along which shear displacement has occurred, provide especially favorable sites for gradient reactions. Newly opened dilatational cracks attract fluid through flow focusing and allow concentrated, relatively high fluid velocities across isotherms. Sibson (1987) has recognized the importance of these fault structures in mineral deposition, citing examples from the Camp Bird vein system, Colorado, and the Chuquicamata copper deposit in Chile. Sibson, Robert and Poulsen (1988) likewise associate the formation of many mesothermal gold–quartz deposits in high-angle fault veins to fluid discharges following fracture.

In summary, in the gradient reaction scenario, rising interstitial fluid becomes uniformly supersaturated and downwardly infiltrating fluid uniformly unsaturated relative to fluid in equilibrium with the suite of minerals in the surrounding fabric. As time goes by, the relative abundance of these minerals changes, some increasing and some decreasing as the reaction continues. The process is ubiquitous, occurring throughout those parts of the fabric where the fluid flow crosses isotherms or isobars. In a fracture–matrix medium, the interstitial fluid velocities are much greater in the fractures than in the matrix blocks so that, from (5.83), the rate of mineralization is concentrated there. The original minerals in the blocks may persist in the blocks for times of the order of the exchange time E^{-1}, proportional to the sizes of individual blocks. As long as all the minerals are present, the interstitial solution is buffered close to this equilibrium value for the ambient temperature and pressure. However, if locally, one of the minerals involved in the reaction say S_1 disappears entirely, the interstitial fluid there moves to a new equilibrium among the remainder of the minerals present. As this fluid moves into an adjacent region where S_1 is not yet depleted, a reaction front may develop between the now distinct mineralogical regions and start to propagate as described in the preceding section (see also Hewett, 1986).

5.5.3 Enhancement and destruction of porosity

These reactions gradually lead to changes in porosity and consequently, in permeability. Even if the driving forces are maintained, the flow patterns evolve as the distributions of permeability change. In simple cementation, the flow is faster in regions of high permeability, so that if the equilibration length l_E is small compared with the flow dimensions, the rate of cementation is greater there and the porosity and permeability decrease more rapidly than in regions where they are already low and the flow slower. Cementation, therefore, tends to make the porosity more uniform throughout the region as the permeability and porosity gradually decrease with time – the deposit is disseminated. With dissolution (e.g. of carbonate rocks), the reverse occurs. Regions of initially relatively high porosity and permeability attract flow focusing, with enhanced dissolution producing enlargement of the pores, further local increase in the permeability, and ever-increasing flow. Merino, Ortoleva and Strickholm (1983), Merino (1984), Chadham *et al.* (1986), and Ortoleva *et al.* (1987a) have discussed this kind of geochemical self-patterning or infiltrative instability in considerable detail. Ortoleva *et al.* (1987b) ascribe many instances of textural patterns such as cement bands, metamorphic layering, and oscillatory zoning to the process and give a linear instability analysis showing that a plane dissolution front is indeed unstable, small perturbations to the front developing into long fingers.

Replacement reactions, such as the conversion of limestone to dolomite, are also generally associated with volume changes that can be of either sign. The dolomite reaction

$$2CaCO_3 + Mg^{2+} \Leftrightarrow CaMg(CO_3)_2 + Ca^{2+}$$

is accompanied by about a 15% decrease in volume of the solid phase, with consequent enhancement in porosity, increase in local flow speed, and increased reaction rate. It is natural to expect, then, that the distribution of dolomite deposited into a limestone matrix by this reaction will occur bi-modally, with dolomite layers interbedded between layers of unaltered limestone. This kind of structure is indeed found in atolls and in once-submerged carbonate banks such as the Latemar Massif in the Dolomites (Wilson, Hardie and Phillips, 1990). The same general principle applies to ores or other minerals deposited by reaction in a permeable or fracture-matrix flow environment. If the mineral formation involves a decrease in volume of the solid, the mineral will generally be deposited in veins or lodes; if it involves an increase in volume of the solid, the deposit is generally disseminated.

An estimate of the rates at which these physical alterations proceed can be found simply, as shown by Wood and Hewett (1982) and Lichtner (1985). Let Q represent the rate of addition of species to the solution (mass per unit volume of the fabric per unit time). Suppose that mineral with density ρ_1 is replaced mole for mole by another with density ρ_2. The rate of disappearance of mineral 1 into solution results in a rate of increase in the porosity

$$\rho_1 \frac{\partial \phi}{\partial t} = m_1 Q, \tag{5.89}$$

where m_1 is the molecular weight of this mineral. The deposition of mineral 2 leads to a rate of decrease in porosity

$$\rho_2 \frac{\partial \phi}{\partial t} = -m_2 Q, \tag{5.90}$$

so that the net rate of change is

$$\frac{\partial \phi}{\partial t} = \left(\frac{m_1}{\rho_1} - \frac{m_2}{\rho_2} \right) Q. \tag{5.91}$$

The initial rate of change in porosity in a single-mineral fabric gives a measure of the time scale in these processes. If the initial porosity is designated as ϕ_0, then from (5.89) its rate of change with time is given by

$$\frac{\partial}{\partial t} \left(\frac{\phi}{\phi_0} \right) = \frac{mQ}{\rho \phi_0} \tag{5.92}$$

with $Q > 0$ for dissolution and $Q < 0$ for deposition. The time scale for fabric alteration can be taken as

$$\tau = \rho \phi_0 / m Q, \tag{5.93}$$

where ρ is the density of the mineral undergoing alteration, m its molecular weight, and Q represents the rate of dissolution (moles per unit volume of the fabric per unit time), This geological time scale can be quite short, of order 10^6–10^7 years.

Boles and Ramseyer (1987) have described an example of very rapid *carbonate* cementation in the San Joaquin Basin of California. The extraction of gas appears to have produced in the limestone such a rapid interstitial flow of isotopically light water that a dolomite cement, rich in iron, has precipitated within the past twenty years! McBride (1989) reviewed quartz cementation in sandstones, and another study documenting calcite and iron sulfide cementation along the margins of salt domes is that of McManus and Hanor (1988).

Over time intervals of order τ or longer, the changing porosity modifies the flow field. The relationship between permeability and porosity depends on the geometry of the fluid pathways. Generally, from Section 2.4, $k \sim 10^{-2}\phi\delta^2$, where δ is the characteristic diameter of the fluid pathways. If they are predominantly tubular, then from (2.2), $\phi \sim n\delta^2$ where n is the number of tubes per unit cross-sectional area, so that as the pore diameter changes, $k/k_0 = (\phi/\phi_0)^2$ where k_0 and ϕ_0 are initial values. If, however, the fluid flows along an intersecting network of cracks or fissures, $\phi \sim \lambda\delta$ from (2.3), where λ is the length of crack per unit cross-sectional area and so $k/k_0 = (\phi/\phi_0)^3$. More generally, it might be asserted that

$$\frac{k}{k_0} = \left(\frac{\phi}{\phi_0}\right)^n,$$

where the power n lies between 2 and 3.

If both the pressure gradient and temperature field are maintained, the flow velocity u is proportional to k and therefore to ϕ^p. In a gradient reaction, the rate of dissolution or mineralization per unit volume is also proportional to the infiltrating fluid velocity,

$$(Q(t)/Q(0)) - (u(t)/u(0)) = (\phi(t)/\phi(0))^n \tag{5.94}$$

and (5.92) becomes

$$\frac{d}{dt}\left(\frac{\phi}{\phi_0}\right) = \frac{1}{\tau}\left(\frac{\psi}{\phi_0}\right)^n, \tag{5.95}$$

where the fabric alteration time scale τ of (5.93) is positive for dissolution and is to be taken negative for cementation. Since $p \neq 1$, the solution to (5.95) is

$$\frac{\phi}{\phi_0} = \frac{1}{\{1 - (n-1)(t/\tau)\}^{1/(n-1)}}, \tag{5.96}$$

which is illustrated in Figure 5.14.

When Q and therefore τ is negative, cementation proceeds ever more slowly as the pores clog. The porosity is reduced to half its initial value in a time interval between τ and 3τ, to one-tenth between 9τ and 50τ according to the geometry of the structure. In dissolution conditions, with τ positive, however, a calamity occurs when $t = \tau/(1 - p)$ (that is, between $\tau/2$ and τ) since $\phi(t) \to \infty$! *Within a finite time*, the porosity increases dramatically; in reality, the fabric may collapse.

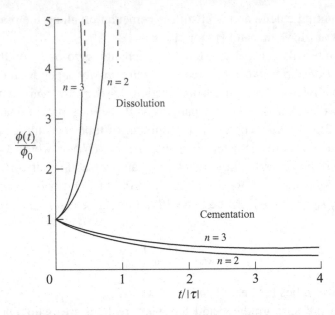

Figure 5.14. Variations in porosity in a flow under conditions of dissolution and cementation, with a maintained pressure gradient. The time scale τ is given by (5.93). Fabrics with primarily tubular interstices have $n \sim 2$; if they are in the form of cracks or fissures, $n \sim 3$.

The time scales for replacement reactions can be found similarly. From equation (5.91)

$$\left(\frac{m_1\rho_2 - m_2\rho_1}{\rho_2}\right)^{-1} \frac{\partial\phi}{\partial t} = \frac{Q}{\rho_1}, \tag{5.97}$$

which is identical to (5.95) except for the additional modifying factor on the left, which simply changes the time scale but does not alter the structure of the solution (5.96), illustrated in Figure 5.14. The time scale for fabric alteration is then somewhat longer:

$$\tau = \frac{\rho_1\phi_0}{m_1 Q(0)} \left\{1 - \frac{m_2\rho_1}{m_1\rho_2}\right\}^{-1} = \frac{\phi_0}{Q(0)} \left\{\frac{\rho_1\rho_2}{m_1\rho_2 - m_2\rho_1}\right\}. \tag{5.98}$$

Note the singularity in this expression when $(m_1\rho_2 - m_2\rho_1) = 0$, which makes sense mathematically, but not physically. It derives from (5.97), which is singular when the deposition and dissolution rates are exactly equal and $\partial\phi/\partial t = 0$.

A number of authors have documented the reduction of porosity by physical compaction. Wilson and McBride (1988) analysed the variations in porosity with depth in sandstones of the Ventura Basin, California, those most deeply buried having lost a total absolute porosity of 26% from a presumed initial value of 40%, or 0.4. In compaction, the expelled interstitial fluid moves generally upward relative to the matrix, though if the compaction is the result of continued sedimentation and the rate of sedimentation is greater than the interstitial velocity produced by compaction, the depth of individual fluid elements below the depositional interface may increase. Bonham (1980) has explored the consequences of this effect on the migration of hydrocarbons.

5.6 The mixing zone scenario

Alterations in mineral composition or precipitation or dissolution can also occur by the mixing of different interstitial waters. One common example is provided by a coastal salt wedge, where sea-water infiltrates beneath the freshwater outflow of a coastal aquifer and mixing occurs across the interface. Fluid dispersion is necessarily involved, and this is greatly augmented by the presence of lenses or more permeable layers, as described in Section 3.4. If one is concerned with the structural details of the reaction patterns, it is necessary to resolve the flow details on the scale of the lenses, but if one is content with a description of only the general characteristics of the region, it may be adequate to represent the mixing in terms of the dispersivity α_D. It should be remembered, though, that this is a fairly coarse approximation. The overall mixing behavior can be represented only asymptotically in terms of a dispersion coefficient when the lens scale is small compared with the scale of the overall flow, and this may not be so in many applications. Usually, however, the lens structures or the distributions of permeability variations are not known in detail, and this approximation may be the best available.

In any event, when the equilibration length l_F is small, the source term is again specified by (5.81), but now the length scale over which the fluid composition varies may be small, of the order of the lens thickness or the local dispersivity. The equilibrium concentration c_E of any particular species may depend on the local mixing ratio of the two fluids, and it will be seen that reactions occur when c_E is other than a linear function of this ratio.

For the sake of definiteness, consider the dissolution or precipitation of calcium carbonate in a coastal freshwater, salt-water mixing zone. The equilibrium concentration of dissolved $CaCO_3$ in the freshwater of the aquifer is generally less than that in sea-water, and although the concentration of dissolved $NaCl$ may not enter the dissolution process directly, the salinity S is a useful index of the mixing ratio. The equilibrium concentration c_E is therefore a function of S (Figure 5.15). When sea-water and freshwater, each initially saturated with respect to $CaCO_3$, are mixed, the resulting concentration of $CaCO_3$ lies along the straight line joining $(c_E)_s$ and $(c_E)_f$ in this figure. Unless $c_E(S)$ is precisely linear in S, the mixture may be unsaturated or supersaturated over the whole range of the mixing ratio S if the curvature

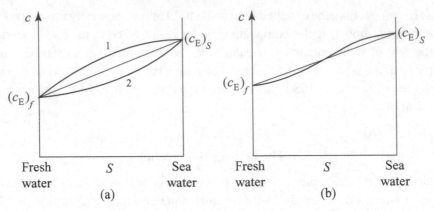

Figure 5.15. When sea-water and freshwater, both saturated with respect to $CaCO_3$, are mixed, the resulting concentration lies along the straight line joining the fresh and salt water equilibrium concentrations $(c_E)_f$ and $(c_E)_s$. The salinity S (of dissolved NaCl) may not influence the dissolution process, but it does provide a useful index of the mixing ratio. If $c_E(S)$ lies above the straight line in (a), $\partial^2 c_E/\partial s^2 < 0$ (curve 1), the solution is unsaturated with respect to $CaCO_3$ and dissolution occurs. If it is below, as in curve 2, the reverse occurs. If the curvature changes sign as in (b), dissolution occurs at high salinities and deposition at low.

of $c_E(S)$ is everywhere concave downward or upward (Figure 5.15a) or unsaturated over parts of the range and supersaturated over others (Figure 5.15b). When the solution is unsaturated, we would expect local dissolution; when supersaturated, precipitation.

The rate of dissolution of $CaCO_3$ per unit volume of fluid is specified by (5.85) with the additional dispersion term

$$Q_C = \overline{\mathbf{v}} \cdot \nabla c_E - \overline{v}\alpha_D \nabla^2 c_E, \qquad (5.99)$$

where now $c_E = c_E(S)$ and $\overline{v}$ is the mean interstitial flow speed. The distribution of salinity S is governed by the steady advection–diffusion balance (2.52) but without a source term:

$$0 = \overline{\mathbf{v}} \cdot \nabla S - \overline{v}\alpha_D \nabla^2 S. \qquad (5.100)$$

Now, since $c_E = c_E(S)$,

$$\nabla c_E = \left(\frac{\partial c_E}{\partial S}\right) \nabla S$$

and

$$\nabla^2 c_E = \left(\frac{\partial^2 c_E}{\partial S^2}\right)(\nabla S)^2 + \left(\frac{\partial c_E}{\partial S}\right)\nabla^2 S,$$

from the chain rule for partial differentiation. Consequently,

$$
Q_C = \left(\frac{\partial c_E}{\partial S}\right)\{\bar{\mathbf{v}} \cdot \nabla S - \bar{v}\alpha_D \nabla^2 S\} - \bar{v}\alpha_D \left(\frac{\partial^2 c_E}{\partial S^2}\right)(\nabla S)^2,
$$

$$
= -\bar{v}\alpha_D \left(\frac{\partial^2 c_E}{\partial S^2}\right)(\nabla S)^2, \tag{5.101}
$$

in virtue of (5.100).

This very simple and important result shows explicit dependence on the mixing parameter and also (a) that the rate of dissolution when $(\partial^2 c_E/\partial S^2) < 0$, or precipitation when $(\partial^2 c_E/\partial S^2) > 0$, is proportional to the curvature of the function $c_E(S)$, which is determined by the solution kinetics; (b) that the rates are proportional to the *square* of the salinity gradient and are therefore greatest when this gradient is largest, at the sea-level outflow in the salt wedge situation; and (c) that the rates increase linearly with the interstitial flow velocity. The formation of sea-level caverns in the limestone Yucatan aquifer can possibly be interpreted in these terms.

More generally, when fluid in an aquifer is drawn into a more permeable lenticular inclusion as described in Section 3.4, the interstitial flow speed is amplified by the focusing ratio G, as is any vertical gradient in interstitial fluid concentration, so that the curvature $\partial^2 c_E/\partial S^2$ increases as G^2. The reaction rate, then, from (5.101) increases by the factor G^3, which can be very large. If $(\partial^2 c_E/\partial S^2) \neq 0$, mixing zone reactions *are especially concentrated* in the most permeable layers – the rate of gradient reactions and the speed of advance of isothermal reaction fronts increase only linearly with the flow velocity.

5.7 Isotherm-following reactions

Many metamorphic reactions are not driven by interstitial fluid flow but by the history of temperature variations of a mineral assemblage, following either volcanic intrusion or tectonic events that force the matrix to a greater depth. Water may be *produced* by the reaction and subsequently move through the matrix; in these dehydration reactions

$$
S_1 \Leftrightarrow S_2 + H_2O. \tag{5.102}
$$

An example is the dehydration of muscovite (white mica):

$$
\underset{\text{Muscovite}}{KAl_3Si_3O_{10}(OH)_2} + \underset{\text{Quartz}}{SiO_2} \Leftrightarrow \underset{\text{K-feldspar}}{KAlSi_3O_8} + \underset{\text{Sillimanite}}{Al_2SiO_5} + \underset{\text{water}}{H_2O}, \tag{5.103}
$$

where S_1 and S_2 of the preceding equation represent the assemblages on the left and right, respectively. The equilibrium diagram for this reaction is of the general form illustrated in Figure 5.16, the equilibrium temperature being almost independent

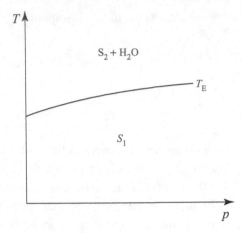

Figure 5.16. The equilibrium diagram for a dehydration reaction with a discrete $T_E(p)$, almost independent of pressure; contrast this with Figure 5.17.

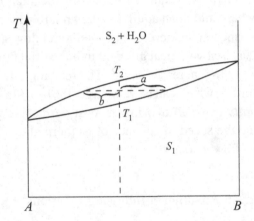

Figure 5.17. The equilibrium diagram for dehydration reactions in low-grade metamorphosis, with a range of transition temperatures.

of pressure, so that as the temperature of a block of the assembly S_1 increases, it remains stable until its temperature exceeds the equilibrium value T_E and the reaction moves to the right. In spatial terms, the isotherms move through the fabric and the reaction zone follows behind the moving equilibrium isotherm.

In low-grade metamorphism, which usually involves complex assemblages and a dense sequence of reactions, the equilibrium line may be replaced by a transition zone (Figure 5.17). At a given ratio A/B of major constituents in S_1, reaction commences when the temperature exceeds T_1 and is complete when it exceeds T_2. At intermediate temperatures, the assemblage containing both S_1 and S_2 is in local equilibrium, the proportion of S_1 being a/b in Figure 5.17 or approximately

$(T - T_1)/(T_2 - T_1)$, so that as the temperature gradually increases – in a dehydration reaction, for example – water is gradually released. In terms of the spatial distribution of fluid generation, for the case illustrated in Figure 5.16, the source zone will be found to be limited to a restricted region following the equilibrium isotherm, whereas for that of Figure 5.17, it is more diffuse.

In either case, if water is produced by the reaction, the interstitial fluid pressure increases. If the generation rate of fluids is sufficiently rapid and the vertical permeability in the sequence above the reaction zone is sufficiently small to form a seal or cap, the interstitial fluid pressure below the cap may be significantly above hydrostatic, providing a large vertical pressure gradient and a scenario for local rock fracture, chemically induced. If the rock permeability above the reaction zone is initially relatively large, the excess pressure can be released by upward fluid migration, although this process tends to be self-sealing. The rising interstitial fluids contain solutes in approximate equilibrium with their surroundings, and if the equilibrium concentration decreases with temperature and pressure (as is the case for dissolved silica), mineral deposition will occur in accordance with equation (5.87). The porosity and permeability of the rock above the reaction zone will therefore decrease in time, increasing the vertical pressure gradient if the fluid generation rate remains even approximately constant. Under appropriate conditions and after sufficient time, local fracture may occur in the less permeable layers. This temporarily relieves the excess pressure until further deposition in veins along the fractures reseals the matrix and the cycle recurs. The same cycles of events can occur when the fracturing is induced seismically, as Sibson et al. (1988) indicate, a sequence they call "seismic pumping."

It is of interest to establish the general conditions necessary for the buildup of significant overpressure in the interstitial fluids and for the occurrence of chemically driven rock fractures. This dynamical process requires the movement of isotherms through the matrix, as may occur during and following active overthrusting. How long does it take to develop pressures that approach the lithostatic value, where relative to the isotherms does this occur first, and how long do the pressures persist once the tectonic processes have become inactive?

5.7.1 The reaction zone

When the isotherms move through the matrix, the general pattern of the reaction is the same whether or not dehydration is involved. Let s be the number of moles of solid veactant S_1 per unit volume. In the reaction, this decreases from its initial value s_0 to zero as S_1 is replaced by S_2.

$$\frac{ds}{dt} = -Q_s. \tag{5.104}$$

In the situation illustrated in Figure 5.16, when the temperature T is less than the equilibrium temperature T_E there is no reaction and $Q_S = 0$, but when $T > T_E$ the rate of disappearance of S_1 is proportional to the amount per unit volume present and, near equilibrium, to the temperature difference $T - T_E$. Accordingly, Q_S can be represented as $\gamma s(T - T_E)/T_E$ where the kinetic reaction rate γ has dimensions of (time)$^{-1}$. Thus,

$$\frac{ds}{dt} = 0, \qquad T < T_E,$$
$$= -\gamma s(T - T_E)/T_E \qquad T > T_E. \tag{5.105}$$

In volcanically active regions the equilibrium isotherm can advance vertically through the mineral assemblage with a speed W_E of order centimeters or meters per year but in regions that are subsiding because of accretion above, W_E may be smaller by a factor of order 10^{-5}. Nevertheless, at a point near and beneath the level of the equilibrium isotherm and fixed with respect to the medium,

$$T - T_E = (GW_E)t, \tag{5.106}$$

where $G = -\partial T/\partial z$ is the local geothermal gradient and t is the time interval since the isotherm passed that point. Alternatively, the speed of advance is

$$W_E = (\partial T/\partial t)/G, \tag{5.107}$$

i.e. the rate of change of temperature at a fixed point divided by the local geothermal temperature gradient. Thus, behind (and below) the equilibrium isotherm,

$$\frac{\partial s}{\partial t} = -\gamma W_E t \left(\frac{G}{T_E}\right) s < 0,$$

while ahead of and above it, no reaction has yet occurred and $s = 0$. In terms of vertical distance $Z = W_E t$ behind the rising equilibrium isotherm,

$$W_E \frac{\partial s}{\partial Z} = -\left(\frac{\gamma G}{T_E}\right) Zs \qquad Z > 0, \qquad \text{below,}$$
$$= 0 \qquad \text{above.} \tag{5.108}$$

Ahead of and above the reaction zone, $s = s_0$, so that the distribution of the reactant S_1 is

$$s = s_0 \exp -(Z/l_S)^2 \qquad \text{below}$$
$$= s_0 \qquad \text{above} \tag{5.109}$$

where the thickness scale of the reaction zone

$$l_S = \left(\frac{W_E T_E}{2\gamma G}\right)^{1/2} \tag{5.110}$$

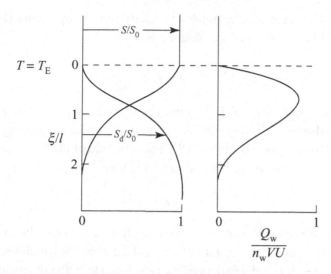

Figure 5.18. The dehydration reaction zone behind the equilibrium isotherm that is moving vertically upward through the matrix with speed $U = W_E$. The fraction of the hydrate S/S_0 is equal to 1 above, or ahead of, the reaction zone and zero behind it. The fraction of the dehydrate is its mirror image, and the thickness scale of the reaction zone is given by equation (5.110).

These distributions are illustrated in Figure 5.18. When the speed of advance W_E of the equilibrium isotherm is small, as in geothermally inactive regions, the thickness of the reaction zone decreases and the boundary between the assemblages S_1 and S_2 approaches a discontinuity.

In lower-grade metamorphism (Figure 5.17) the thickness scale of the reaction zone is greater, since the sequence of reactions occurs over the finite temperature range (T_1, T_2). Regions of contact metamorphism surrounding magmatic intrusions frequently show gradations in mineral assemblages rather than sharp contacts, reflecting this sequence of reactions, or possibly an episode of rapid frontal advance.

5.7.2 Dehydration

If the reaction involves dehydration, with 1 mole of the mineral or mineral assemblage S_1 releasing n_W moles of water, the volumetric rate of fluid generation $Q_W = n_W V Q_S$, where V is the molar volume of the fluid at the ambient pressure. Thus,

$$Q_W = n_W V Q_S = -n_W V \frac{ds}{dt}, \qquad (5.111)$$

from (5.104), and in the case of a discrete equilibrium temperature (Figure 5.16), the rate of fluid generation per unit volume is

$$Q_W = -n_W V W_E \frac{\partial s}{\partial Z}, \qquad (5.112)$$

which is positive since $\partial s/\partial Z$ is negative, as specified by (5.109). This rate is a maximum at a distance $2^{1/2} l_S$ behind the moving isotherm, decreasing to zero with increasing distance beyond (Figure 5.18). The total rate of fluid generation per unit area of the front is the vertical integral of (5.111), i.e.,

$$\hat{Q}_W = n_W V s_0 W_E \qquad (5.113)$$

which involves the molar density s_0 of the dehydrating mineral, the moles of water released by each of them, the molar volume and the speed at which the equilibrium isotherm advances. It is independent of the reaction rate γ and the local geothermal temperature gradient G. As (5.110) shows, these quantities, together with the speed at which the equilibrium isotherm rises through the assemblage, determine the thickness of the dehydration zone, but not the total amount released.

In lower-grade metamorphism (Figure 5.17), the distance over which the reaction occurs is greater but (5.113) remains valid. This can be seen either by considering the overall process as a sequence of small steps of the kind just described, or alternatively as follows. At a temperature T intermediate between T_1 and T_2, the moles of water that can still be produced by dehydration is

$$n_W s_0 \left(\frac{T_2 - T}{T_2 - T_1} \right).$$

As the temperature increases at each point, the volumetric rate of fluid generation is minus the time derivative of this, above, times the molar volume V, or

$$\frac{n_W s_0 V}{(T_2 - T_1)} \frac{\partial T}{\partial t}.$$

If, again, the isotherms move with speed W_E through the assemblage, the rate of generation is therefore approximately uniform with distance between the isotherms T_1 and T_2 and equal to

$$\frac{n_W s_0 V}{(T_2 - T_1)} W_E G \qquad (5.114)$$

where $G(> 0)$ is the local temperature gradient. The distance between these two isotherms is approximately $(T_2 - T_1)/G$. The total rate of fluid generation per unit area of the zone is the product of $(T_2 - T_1)/G$ and (5.112), which recovers (5.113).

5.8 Paleo-convection and dolomite formation in the Latemar Massif

A particularly striking and well-documented example of the consequences of paleo-convection is provided by the intensity and distribution of dolomite formed in the Middle Triassic Latemar Buildup of the Dolomites in northern Italy. The Latemar Massif itself is a small, isolated carbonate platform, about 5 km in diameter and some 700 m thick, that preserves a stratigraphic record of shallow marine carbonate deposition from the middle Triassic (*c.* 200 million years ago) in its vertically stacked "layer cake" deposits. Preferential erosion has revealed a remarkable exposure that has been studied extensively by Bosellini and Rossi (1974), Gaetani *et al.* (1981), Bosellini (1984), Goldhammer (1987), Wilson *et al.* (1990), and others. The Latemar Buildup is unusual in its proximity to the Predazzo volcanic–intrusive complex that was active late in its formation, at which time lava flows filled basins adjacent to the Latemar and a ring–dike complex intruded into it.

The stratigraphy of the formation is evident (to the trained eye) in the remarkable cliff exposures. The lower portion, consisting of approximately 300 m of horizontally bedded grainstones (averaging about 0.5 m in thickness), exhibits abundant early calcite cementation and represents periods of shallow-water subtidal deposition punctuated by intervals of submarine cementation. Gaetani *et al.* (1981) refer to the lower Middle Triassic portion as the "Lower Edifice." Above it, the upper platform consists of some 400 m of horizontal cyclically bedded carbonates containing evidence of repeated sub-aerial exposure (Hardie, Bosellini and Goldhammer, 1986; Goldhammer, Dunn and Hardie, 1987) alternating with shallow submergences. At the top of each cycle there is a thin (5–15 cm) layer containing dolomite with evidence of incipient soil formation. Cross-cutting these layers are small dispersed veinlets of dolomite cement, but the total volume of dolomite in these microstructures is insignificant. There is almost no lateral variation in lithography in the interior of the Latemar Platform until, at its edges, it is bounded by steeply sloping foreslope breccias and grainstones deposited as the platform aggregated.

The original limestone of the formation has been massively but incompletely converted to dolomite in a zone restricted to the core of the massif, and having well-defined boundaries with the unconverted limestone. It is possible to map the geometry of the dolomite body that cross-cuts the primary bedding and fans out toward the top. A total of 2–3 km^3 of massive porous dolomite has been generated over a time interval estimated to be a few million years (certainly not more than 10 or 20) – this represents roughly 2×10^{13} moles of dolomite. The most credible source of the magnesium required for the conversion of limestone to dolomite is the sea-water that surrounded the platform. Assuming an initial magnesium concentration in the sea-water of 5.3×10^{-2} molal (Garrels and Thompson, 1962), Wilson *et al.* (1990) calculate that about 500 km^3 of sea-water would be required to furnish the necessary amount of magnesium if all of it were available for conversion to dolomite – several hundred times the total pore volume of the Latemar. A more refined calculation based on the amount of magnesium present in excess of the Mg^{2+}/Ca^{2+} ratio at calcite–dolomite equilibrium leads to a rather larger value of sea-water needed – some 1000 km^3 at an internal temperature in the massif of 100 °C. Quite evidently, a continuing circulation of sea-water from the surroundings, through the platform and continuing over a considerable time period, is required to account for the extent of this replacement.

The existence of the volcanic intrusions that occurred after the formation of the platform indicates the presence at the time of elevated temperatures near the basement that would drive a convective circulation in the permeable bank of one kind or another, as described in Chapter 4. The questions of interest are (i) whether the pattern of circulation, as evidenced by the pattern of the massively dolomitized region, can unambiguously indicate the type of flow-controlled reaction involved, (ii) whether the pattern points to values of the Rayleigh number and flow speed during convection that are consistent with the overall geometrical and geological constraints, and (iii) whether, with these values, the convection can deliver sufficient Mg^{2+} ions to provide for the extent of the dolomite formation in a geologically reasonable time interval, a few million years.

Figure 5.19 is a simplified map given by Wilson (1989) of the present horizontal extent of the dolomitized region in relation to the depositional facies of the Latemar Buildup (omitting dikes and volcanic breccias) indicating its generally centered configuration in the larger limestone platform. Figure 5.20 gives the distribution of intense dolomitization projected onto a cross section showing the overall mushroom-shaped distribution. At the base, massive dolomite is in vertical contact with the limestone and a core of brecciated dolomite. This grades upward into a zone of large, closely spaced pods of massive dolomites, which become more widely spaced in the mushroom cap. On a smaller scale, it is seen in Figure 5.21 that while the dolomite pods clearly cross-cut the depositional bedding, fingers of dolomite reach out along more permeable bedding planes, the transition surfaces from limestone to dolomite remaining sharp. These characteristics indicate clearly that the reactions occurred in isothermal reaction fronts (Section 5.4) that went to completion and moved through the fabric at speeds proportional to, but much less than, the local flow speeds and, consequently, proportional to the local permeability. In some places, these left behind pockets of less permeable, unaltered limestone. The pattern is quite inconsistent with what would be produced by gradient reactions alone, in which the reactions are more diffuse and ubiquitous, though more intense in high-flow regions than in those where the flow is slow. The more generalized and

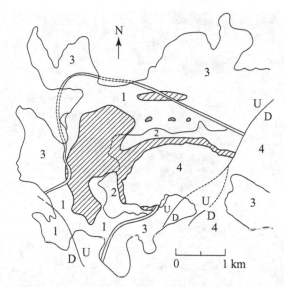

Figure 5.19. A schematic geological map of the Latemar Buildup near Predazzo in the Italian Dolomite region, after Wilson (1989). The dolomite region is cross-hatched. Key: 1, the Ladinian Platform of Latemar limestone, whose margin is indicated by the double line; 2, the Lower Edifice; 3, foreslopes (Marmolada limestone); 4, volcanics. Faults are indicated by the single lines, with the direction of relative displacement (up, down).

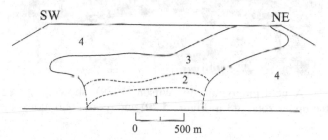

Figure 5.20. The intensity of dolomite formation in the Latemar buildup, projected onto a cross section, after Wilson (1989). In region 1, massive dolomite is in vertical contact with limestone having a core of brecciated dolomite. Region 2 contains large, closely spaced pods of massive dolomite, grading into region 3, above, where the pods are more widely spaced. Region 4 contains limestone, with only sparse dolomite.

dispersed centimeter-scale veinlets noted earlier *are* probably the result of gradient reactions.

Isothermal reaction fronts follow the streamlines, so that the pattern of massive dolomitization indicates the pattern of the ancient flow – a central rising core of fluid fanning out toward the top. The general mushroom shape of the configuration

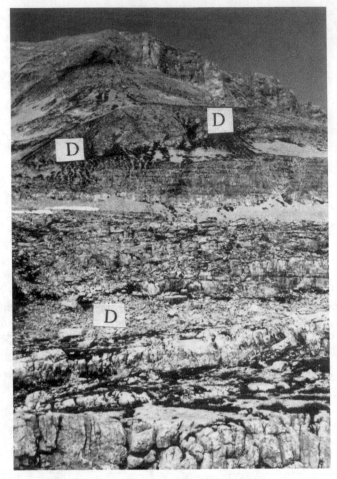

Figure 5.21. A field photograph by Wilson (1989), showing sparse dolomite lenses in the upper zone (#3 in Figure 5.20) and more continuous layers below (D).

is evidence that the circulating fluid did not generally move freely across the upper surface of the platform and is consistent with the geological indications of repeated subaerial exposure.

More quantitative estimates of the flow characteristics can be made by the use of the scaling relations of Section 4.5 together with the measured geometry of the massively dolomitized region. The total platform is about 700 m thick. In this ancient flow, the thickness of the inaccessible lower boundary layer δ was close to that of the cap, which in the Latemar is about 300 m. The stem diameter $2R_S$ shown in Figure 5.20 is about 1300 m and the radius R_P of the platform is some 2500 m. With these values, the Rayleigh number of the paleo-convection, given by

(4.58), was

$$Ra_V \sim \left(\frac{R_P}{R_S}\right)^2 \frac{h}{\delta} \sim 35. \tag{5.115}$$

This value is certainly sufficient to produce significant advective distortion of the isotherms and the warm mushroom-shaped structure described in the previous section. From (4.55), the dimensions above and a thermal diffusivity $\kappa \approx 6 \times 10^{-7}$ m^2/s, the vertical transport velocity up the stem is found to be

$$w_0 \sim 3 \times 10^{-8}\,\text{m/s} \sim 1\,\text{m/yr}.$$

The cross-sectional area of the stem was close to 1.5 km^2, so that the volume flux was about 1.5×10^{-3} km^3/yr, so that in one million years, about 1500 km^3 would have passed through the system, which is surprisingly consistent with the estimate made by Wilson *et al.* (1990) from geochemical evidence.

   If the primary deposits of dolomite have resulted from the passage of reaction fronts, the front propagation speed U_F referred to the fluid transport velocity $u_0 \sim w_0$ is given by (5.53). In that equation,

$$\frac{U_F}{w_0} \approx \frac{U_F}{u_0} = \frac{n_S(c_E - c_0)}{s_0},$$

$n_S = 1$ (one mole of product per mole of solute) and $c_0 = 0$ (negligible Mg^{2+} ahead of the reaction front). Consequently, the ratio of the speed of advance of the front to the fluid transport velocity is equal to the ratio of solute concentration entering the front to the concentration of solid Mg^{2+} that the front leaves behind, as illustrated in Figure 5.7. Since the concentrations enter as a ratio, any units can be used, provided they are consistent. The concentration c_E of Mg^{2+} in sea water is 1.3×10^{-3} kg/l (Weast, 1972). Dolomite has a relative density of 2.8 and a molecular weight of 184, of which 24 represents Mg^{2+}, so that the mass of magnesium per liter of dolomite is

$$s_0 = 2.8 \times (24/184) = 0.36\ \text{kg/l}.$$

Thus

$$U_F \approx w_0(c_E/s_0) = 1 \times (1.3 \times 10^{-3}/0.36) = 3.6 \times 10^{-3}\ \text{m/yr},$$

about 3.6 millimeters per year or 3.6 kilometers in one million years. Although approximately equal to the flow path length in the Latemar Massif, this is a lower limit which presupposes that the matrix is uniform and un-fractured. Bosellini and Rossi (1974) and Wilson (1989) describe it as extensively layered and fractured, where the fluid transport can be expected to occur most rapidly in localized conduits that thread through, and feed into the matrix blocks.

The Rayleigh number for the convection and this total transport rate can be found fairly accurately from the geometry of the massive dolomite formation, but the temperatures cannot, since T_0 occurs in Ra in combination with k_V, whose explicit value is much more uncertain than is the temperature. However, since from (5.115),

$$Ra = \frac{\alpha g T_0 k_V h}{\nu \kappa} \sim 35,$$

then

$$T_0 k_V \sim 35 \frac{\kappa \nu}{\alpha g h} \sim 1.2 \times 10^{-13}\, \text{m}^2 \text{K},$$

with $\nu \approx 0.12 \times 10^{-6}\, \text{m}^2/\text{s}$ and $\alpha \approx 3 \times 10^{-3}\, \text{K}^{-1}$. Fluid inclusion data described by Wilson *et al.* (1990) indicate homogenization temperatures (a lower limit for that at the time of formation) of 75 °C near the top of the formation and 220 °C in the stem. If we take the temperature at the base $T_0 \sim 300$ °C, then $k_V \sim 4 \times 10^{-16}\, \text{m}^2 = 4 \times 10^{-12}\, \text{cm}^2$, which is a reasonable value for a limestone structure such as this. It is important to recognize that in this example, the flow characteristics and the numerical value of the total transport rate were found *without* having to guess the appropriate values of the permeability, which may be uncertain to one or two orders of magnitude. It is therefore gratifying to find that the flow rate found accounts so easily for the transport required in the formation of this massive dolomite body.

5.9 Distributions of mineral alteration in Mississippi Valley-type deposits

Consider now several general examples that illustrate the relationships between mineralization and flow, the first being the distribution of lead–zinc deposits in the Mississippi Valley. The Paleozoic sediments of the greater Mississippi Valley provide the locale for extensive major and minor deposits of galena, sphalerite, barite, and fluorite, all of which possess such striking geochemical and geological similarities among themselves and to those in similar regions on other continents that they are all generally known as Mississippi Valley-type deposits. They have been studied extensively and are the subject of a fine review by Sverjensky (1986). It appears to be well established that they are formed from hot, saline aqueous solutions some time after the lithification of the host rock, but the detailed geochemical processes involved in their formation have been a matter of some dispute (i.e. whether the precipitation reactions involved sulfate reduction or pH changes or dilution, etc.); different possibilities are reflected in differences in the chemically determined coefficients in equations such as (4.25).

Nevertheless, the outstanding characteristics of Mississippi Valley-type deposits, as listed by Sverjensky, include the following.

(i) They occur principally in the limestone or dolostone that forms a relatively thin cover over an igneous or highly metamorphosed basement.
(ii) They consist of bedded replacements, vuggy ores, and veins, the ore being strongly controlled by individual strata; they are generally not associated with igneous rocks.
(iii) They always occur in areas of mild deformation, expressed in brittle fractures, broad domes and basins, and gentle folds.
(iv) The ore is never in the basement rocks, but its distribution is often spatially related to basement highs, with the ore located within sandbanks, ridges, and reef structures that *surround* the basement highs.
(v) The ore is generally at shallow depths, generally less than 600 m relative to the present surface, and was probably never at depths greater than 1200 m.
(vi) Fluid inclusions remaining in sphalerite, fluorite, barite, and calcite always contain dense, saline aqueous fluids and often oil and/or methane.

All these characteristics seem to be consistent with the dissolution of dispersed minerals, fluid transport, and local deposition and concentration, probably by gradient reactions. An important clue is provided by the absence of ore deposits in undeformed regions. Here, any fluid flow is essentially horizontally along the isotherms, $\mathbf{u} \cdot \nabla T \approx 0$ and there is no fabric alteration. In flow through regions of deformation or over basement highs, the flow crosses the isotherms, so that on the flanks of these highs, both $\mathbf{u} \cdot \nabla T$ and w in (5.87) are non-zero. Local deposition occurs in regions of rising fluid when the chemical coefficients in (5.87) are positive or in regions of sinking fluid when they are negative. The fact that they are associated with more permeable regions near basement highs is again consistent with flow focusing there and more rapid deposition in gradient reactions. It does not appear to be known whether the formation of these shallow deposits was in fact associated with fluid discharge regions, where the near-surface temperature gradient would be anomalously large, although Cathles and Smith (1983) argue persuasively that episodic discharges would appear to account for the colorbanding observed in sphalerite (ZnS) in these deposits resulting from variations in iron content.

Nevertheless, the qualitative consistency between the characteristics listed by Sverjensky and those that would be expected from the theory does suggest strongly, though not quantitatively, that gradient reactions associated with basin-wide flow were responsible for the ore deposition. More highly quantitative studies are needed, however, before the association can be regarded as tight. The rate at which minerals were deposited depends on the chemical coefficients involved in equations such as (5.87), but the estimation of these for particular reactions is beyond the scope of this book. Some systems have been considered recently. The well-known work of Garrels and Christ (1965) considers the carbonate system, and Wood (1987b) presents detailed calculations for the calcite–dolomite system saturated either with initially freshwater or with brine having the composition of sea-water. Wood gives

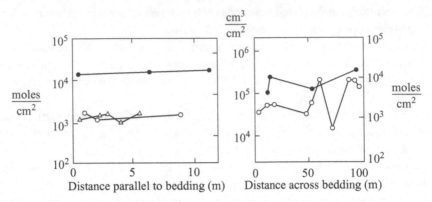

Figure 5.22. Time-integrated fluxes during metamorphism from the limestone member of the Waterville formation, calculated by Baumgartner and Ferry (1991) from Ferry's field samples. The fluxes are constant along the bedding but vary significantly from one bed to another, indicating strong flow channeling.

calculations for both isothermal fronts and gradient reactions (which he terms irreversible and reversible, respectively), taking care to distinguish between them. In either case, however, he finds that the capacity of a sea-water brine to dissolve or produce calcite and dolomite is several hundred times greater than that of initially pure water, so that dolomitization occurs in less time and involves less volume flux when brine is involved.

Finally, the close association between the spatial distribution of metamorphism and paleo-flow has been demonstrated in an important field study by Ferry (1987) on regionally metamorphosed biotite- and garnet-grade rocks of the Vassalboro Formation and the limestone member of the Waterville Formation in south central Maine. The biotite-forming reaction involves decarbonation:

$$\underset{\text{Muscovite}}{KAl_3Si_3O_{10}(OH)_2} + \underset{\text{Ankerite}}{3\,CaMg(CO_3)_2} + \underset{\text{Quartz}}{2\,SiO_2}$$

$$\Leftrightarrow \underset{\text{Biotite}}{KMgAlSi_3O_{10}(OH)_2} + \underset{\text{Plagioclase}}{CaAl_2Si_2O_8} + \underset{\text{Calcite}}{2\,CaCO_3} + 4CO_2.$$

The geometrical distribution of reaction products, with gradual variations through the fabric rather than abrupt transitions, suggests that gradient reactions were involved, rather than the movement of isothermal reaction fronts. Baumgartner and Ferry (1991), interpreting the metamorphism in terms of gradient reactions and using the appropriate stoichiometric coefficients, showed that the total time-integrated volume transports over the period of formation were in the range 10^4–10^6 cm^3/cm^2. Although the variations among different lithological layers were considerable (reflecting variations in permeability), the calculated fluxes from samples at different points in the same layer are very similar (Figure 5.22), indicating the channeling of fluid flow along it. By taking 10^7 years as an upper limit for the time interval involved, Baumgartner and Ferry calculated volume fluxes or transport velocities of order 10^{-2} cm/yr at least.

Another important example of metasomatism (change in the composition of a metamorphic rock) has been discussed by Ferry and Dipple (1992a,b). This involves the alteration of peridotite from the Ivrea zone, northern Italy, with the loss of calcium and magnesium from the rock and the gain of potassium relative to silicon. Thermodynamic calculations indicated that the influence of pressure on the equilibrium concentration c_E in equation (5.87) could be neglected, and the changes in element ratios gave three independent constraints on the two unknown variables, time-integrated volume flux and dissolved chlorine molality (influencing $\partial c_E/\partial T$). They found it possible to choose these two variables so that the three ratios predicted by (5.87) were consistent with those measured. This strongly indicates that gradient reactions were indeed responsible for the alteration.

6

Extensions and examples

6.1 Extensions

The following works supplement and extend the text in various directions. Citations are listed in the References.

Ceazan, M. L., Thurman, E. M. and Smith, R. L. Retardation of ammonium and potassium transport through a contaminated sand and gravel aquifer: the role of cation exchange, 1989. The contamination of groundwater by NH_4^+ and NO_3^-, generally from fertilizer application and the disposal of human and animal wastes, is of growing concern. This paper describes pioneering measurements on the composition and movement of these contaminants in a groundwater plume in south-western Cape Cod.

Cvetkovic, V and Dagan, G. Transport of kinetically sorbing solute by steady random velocity in heterogeneous porous formations, 1994. This is a theoretical study using a Lagrangian approach, different from the one given in this book. The second author has made many important contributions to our understanding of the dispersal characteristics of heterogeneous geological formations.

Fitzgerald, S. D. and Woods, A. W. The instability of a vaporization front in hot porous rock, 1994. Another frontal instability of the general type described in Section 4.9, this one having particular interest in the context of geothermal energy production.

Lyle, S., Huppert, H. E., Hallworth, M., Bickle, M. and Chadwick, A. Axisymmetric gravity currents in a porous medium, 2005. A theoretical and laboratory study with applications to the problems of carbon dioxide sequestration.

Masterson, J. P., Hess, K. M., Walter, D. A. and Leblanc, D. R. Simulated changes in the sources of ground water for public-supply wells, ponds, streams and coastal areas on western Cape Cod, Massachusetts, 2002. This USGS publication is a very readable survey paper illustrating the circulation patterns described in Section 3.2, by some of the leading hydrologists who have worked on the aquifers of Cape Cod.

Tsypkin, G. G. and Woods, A. W. Precipitate formation in a porous rock through evaporation of saline water, 2005. Evaporation fronts of saline water in hot fractures and porous layers involve a variety of processes including advective and diffusive heat transfer, decrease of porosity and sealing, with applications to the natural venting of steam in high temperature geothermal systems.

6.2 Examples

6.2.1 Coastal salt wedges

A freshwater aquifer discharges along the shoreline of a salt lake. Beneath the discharge, saline water from the lake has infiltrated under the fresh water discharge, forming a salt-water wedge above the basement. The volume flux of fresh water per unit length of shoreline Q and the hydraulic conductivity K of the aquifer can be considered constant and the densities of the fresh and saline water are ρ and $\rho + \Delta\rho$. Show that the interface between the fresh and saline regions is at a depth d below the lake water level given by

$$d(x) \approx \left\{ \frac{2\rho}{\Delta\rho} \frac{Qx}{K} \right\}^{1/2},$$

where x is measured inland from a virtual origin some distance offshore. The height of the water table above the lake surface level is smaller than this by a factor $\Delta\rho/\rho$, which is typically about 30.

The fluid mechanics involved here is the same as in (4.84) but the flow geometries differ.

6.2.2 Permeability variations and the rotation vector

The expression

$$\Omega = \nabla \times \boldsymbol{u} = \nabla \left(\ln \frac{k(\boldsymbol{x})}{k_0} \right) \times \boldsymbol{u} \tag{3.9}$$

relating the rotation vector of the transport velocity field can be established and verified in a variety of alternative ways. Most simply but somewhat tediously, take the curl of the Darcy equation in the form $\nabla p = -\nu \boldsymbol{u}/k$. The curl of the gradient vanishes; show by writing out the Cartesian components that the curl of u/k gives (3.9). Alternatively, express Darcy's law in Cartesian tensor notation $\partial p/\partial x_j = -\nu u_j/k$ and take the curl $\varepsilon_{ijk}\partial(..)/\partial x_k$ of both sides.

Testing it: consider a horizontally laminated medium; the gradient of the log-permeability is vertical, and for vertical flow, the pressure gradient is vertical, the right-hand side is zero, no rotation. For horizontal flow, the rotation is perpendicular to the flow

$$\frac{1}{u}\frac{\partial u}{\partial y} = \frac{\partial}{\partial y}\ln\left(\frac{k}{k_0}\right), \quad \text{and so} \quad u \propto k.$$

Draw diagrams.

6.2.3 Confined aquifers

When a permeable aquifer layer is sandwiched between two impermeable strata, there is no flow across the upper and lower bounding surfaces, $z = \zeta_1$ and ζ_2, and the aquifer is said to be confined. Its thickness is characteristically much smaller than its length so that transverse velocities and pressure gradients are much smaller than those along the layer. The volume flux along the aquifer is the integral across the layer of the internal transport velocity:

$$q(x) = (q_X, q_Y) = \int_{\zeta_1}^{\zeta_2} u(x, z)dz,$$

a two-dimensional vector function of position (x, y) in the aquifer 'plane'.

Using the incompressibility condition (2.8) and the boundary condition (2.28) with $\xi = 0$, show (i) that

$$\frac{\partial q_X}{\partial x} + \frac{\partial q_Y}{\partial y} = 0.$$

Therefore (ii) we can define a transport stream function

$$q_X = \partial\psi/\partial y, \qquad q_Y = -\partial\psi/\partial x.$$

(iii) Show also that

$$q = -\mu^{-1}(\overline{k}h)\nabla p,$$

where μ is the fluid viscosity, $\overline{k}$ is the mean permeability, averaged across the aquifer of thickness h. Consequently, the aquifer flow is attracted to more permeable and thicker conduits in the aquifer and avoids less permeable and narrower regions, in accordance with the minimum dissipation theorem.

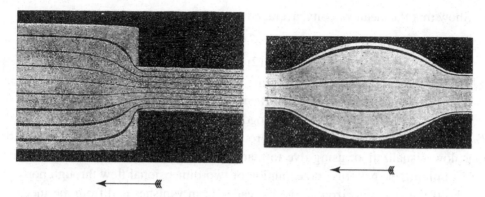

Figure 6.1. Original images from Hele-Shaw (1898) of flow through an abrupt expansion and contraction in an analog of potential or porous medium flow. The image on the right is of "colour bands" (streamlines) in a gradually enlarging and contracting channel and on the left, a sudden enlargement. Note the absence of eddies or circulating flow.

(iv) If the aquifer has a "throat" somewhere along its length where the cross-sectional area or the mean permeability is significantly less than elsewhere, how does this affect the pressure distribution along its length?

6.2.4 An unconfined or surface aquifer with a locally fractured confining layer

An approximately two-dimensional aquifer has a low-permeability retarding layer separating more permeable sub-aquifers above and below. Can the location of a recent fracture zone in the retarding layer significantly influence the degree of confinement? Consider the separate cases of a fracture zone located near the groundwater divide and one close to the discharge region.

6.2.5 The Hele-Shaw cell

One of the most challenging problems in laboratory experimentation on flows in porous media is that of visualizing the patterns of internal streamlines and the onset of instabilities. The flow cell devised by Hele-Shaw (1898) is an extremely useful and accurate analog device that achieves this for two-dimensional pressure or buoyancy driven flows. It consists of two closely spaced, transparent, parallel plates, the gap being filled with viscous fluid. When fluid is introduced around the edges of the cell or when the internal buoyancy varies, motion is generated between the transparent plates. The photographs in Figure 6.1 are from his original report to the magazine *Nature*, which is of historic importance and interesting to read.

Show that the mean velocity, averaged across the gap, $\bar{u}$ is specified by

$$\nabla \cdot \bar{u} = 0, \bar{u} = \frac{\delta^2}{12\nu}\{-\nabla(p/\rho_0) + b\mathbf{l}\}$$

where p is the reduced pressure and b the buoyancy of the fluid in the gap of width δ. (If necessary, see the derivation in Batchelor, 1967).

These equations are precisely analogous to those describing *two-dimensional Darcy flow* in a uniform isotropic porous medium with permeability $\delta^2/12$, so that flow visualization, using dye to mark the streamlines, provides a technique for examining the patterns and evolution of two-dimensional flow through porous media. If the cell is horizontal, the buoyancy term vanishes and from the incompressibility condition, $\nabla^2 p = 0$, as in potential flow. With the cell walls vertical, buoyancy-driven flows can be simulated, using variations in salinity. See the review by Homsy (1987) and papers by Saffman and Taylor (1958), Wooding (1962), (1969) and Huppert and Woods (1995). Thermal convection in a Hele-Shaw cell is complicated by heat losses through the transparent call walls, though Hartline and Lister (1977) showed that corrections are possible.

Finger instabilities in fronts between different fluids in a Hele-Shaw cell have been described by Park and Homsy (1984, 1985). These are dynamically analogous to the Saffman–Taylor instabilities in porous media considered in Section 4.9.

Bibliography

Aagaard, P. and Helgeson, H. C. 1982. Thermodynamic and kinetic constraints on reaction rates among minerals and aqueous solutions, I. Theoretical considerations. *Amer. J. Sci., 282*, 273–285.

Abelin, H. L., Birgersson, J., Gidlund, J. and Neretnieks, I. 1991. Large scale flow and tracer experiment in granite, 1. Experimental design and flow distribution. *Water Resources Res., 27*, 3107–3117.

Adams, J. E. and Rhodes, M. I. 1960. Dolomitization by seepage refluxion. *Bull. Amer. Assoc. Petrol. Geol., 44*, 1912–1920.

Altunel, E. and Hancock, P. L. 1996. Structural attributes of travertine-filled extensional fissures in the Pamukkale Plateau, Western Turkey. *Internat. Geology Rev., 38*, 8, 768–777.

Anderson, G. M. and Burnham, C. W. 1965. The solubility of quartz in supercritical water. *Amer. J. Sci., 263*, 494–511.

Bachman, L. J. 1984. *The Columbia Aquifer of the eastern shore of Maryland, Part 1: Hydrogeology*. Maryland Geological Survey, Report of investigations, No. 40.

Barenblatt, G. I. 1963. On some boundary-value problems for equations of fluid seepage in fractured rocks (in Russian). *Prikl. Mat. Mekh., 27*, 348–350.

Barenblatt, G. I., Entov, V. M. and Ryzhik, V. M. 1990. *Theory of Fluid Flows Through Natural Rocks*. Dordrecht: Kluwer Academic Publishers.

Barenblatt, G. I., Zheltov, Y. P. and Kochina, I. N. 1960. On basic concepts of the theory of homogeneous fluid seepage in fractured rocks (in Russian). *Prikl. Mat. Mekh, 24*, 852–864. (Engl. Transl., *J. Appl. Math. Mech., 24*, 1286–1303.)

Barnes, H. L. (ed.) 1979. *Geochemistry of Hydrothermal Ore Deposits*. New York: Holt, Rinehart & Winston.

Batchelor, G. K. 1949. Diffusion in a field of homogeneous turbulence. *Aust. J. Sci. Res., A2*, 437–450.

Batchelor, G. K. 1953. *The Theory of Homogeneous Turbulence*. Cambridge University Press.

Batchelor, G. K. 1967. *An Introduction to Fluid Mechanics*. Cambridge University Press.

Bathurst, R. G. C. 1975. *Carbonate Sediments and their Diagenesis*. New York: Elsevier.

Baumgartner, L. P. and Ferry, J. M. 1991. A physical model for infiltration-driven mixed volatile reactions and its application to regional metamorphism. *Contrib. to Mineral & Petrol., 106*, 273–285.

Bear, J. 1979. *Dynamics of Fluids in Porous Media*. New York: Elsevier.

Bear, J. and Verruijt, A. 1987. *Modeling Groundwater Flow and Pollution*. Dordrecht: Reidel Publishing Company.

Bird, R. B., Stewart, W. E. and Lightfoot, E. N. 1960. *Transport Phenomena*. New York: Wiley.

Birgersson, L., Moreno, L., Neretnieks, I., Widén, H. and Ågren, T. 1995. A tracer migration experiment in a small fracture zone in granite. *Water Resources Res., 29*, 3867–3878.

Böhlke, J. K. and Denver, J. M. 1995. Combined use of groundwater dating, chemical and isotopic analyses to resolve the history and fate of nitrate contamination in two agricultural watersheds, Atlantic coastal plain, Maryland. *Water Resources Res., 40*, 2319–2339.

Böhlke, J. K., Smith, R. L. and Miller, D. N. 2006. Ammonium transport and reaction in contaminated groundwater: Application of isotope tracers and isotope fractionation studies. *Water Resources Res., 42*, WO5411, doi:10.1029/2005-WR004349.

Boles, J. R. and Ramseyer, K. 1987. Diagenetic carbonate in miocene sandstone reservoir, San Joaquin Basin, California. *Amer. Assoc. Petrol. Geol. Bull., 71*, 1475–1487.

Bonham, L. C. 1980. Migration of hydrocarbons in compacting basins. In *Problems of Petroleum Migration*, pp. 69–88. Amer. Assoc. Petrol. Geol. Studies in Geol. no. 10, Houston.

Bories, S. A. and Combarnous, M. A. 1973. Natural convection in a sloping porous layer. *J. Fluid Mech., 57*, 63–79.

Borkowska-Pawlak, B. and Kordylewski, W. 1985. Cell-pattern sensitivity to box configuration in a saturated porous medium. *J. Fluid Mech., 150*, 169–181.

Bosellini, A. 1984. Progradation geometries of carbonate platforms: examples from the Triassic of the Dolomites, northern Italy. *Sedimentology, 31*, 1–24.

Bosellini, A. and Rossi, D. 1974. Triassic buildup of the Dolomites, northern Italy. In *Reefs in Time and Space*, ed. L. Laporte. Soc. Econ. Paleontol. Mineralog. Special Publ. no. 18, pp. 209–233. Tulsa, Okla.

Boussinesq, J. 1904. Recherches théoretiques sur l'écoulement des nappes d'eau dans le sol. *J. Math. Pures Appl.*, sèr 5, *X*, 5–78, 363–394.

Brace, W. F. 1980. Permeability of crystalline and argillaceous rocks. *Int. J. Rock Mech. Min. Sci., 17*, 241–251.

Caltagirone, J. P. and Bories, S. A. 1985. Solutions and stability criteria of natural convective flow in an inclined porous layer. *J. Fluid Mech., 155*, 267–287.

Carslaw, H. S. and Jaeger, J. C. 1959. *Conduction of heat in solids*. Oxford: Clarendon Press.

Cathles, L. M. 1977. An analysis of the cooling of intrusives by ground-water convection which includes boiling. *Econ. Geol., 72*, 804–826.

Cathles, L. M. and Smith, A. T. 1983. Thermal constraints on the formation of Mississippi Valley-type lead–zinc deposits and their implications for episodic basin dewatering and deposit genesis. *Econ. Geol., 78*, 983–1002.

Ceazan, K. L., Thurman, E. M. and Smith, R. L. 1989. Retardation of ammonium and potassium transport through a contaminated sand and gravel aquifer. The role of cation exchange. *Environ. Sci. Technol., 23*, 1402–1408.

Chadham, J., Hoff, D., Merino, E., Ortoleva, P. and Sen, A. 1986. Reactive infiltration instabilities. *IMA J. Appl. Math., 36*, 207–221.

Chandler, R., Koplik, J., Lerman, K. and Willemsen, J. F. 1982. Capillary displacement and percolation in porous media. *J. Fluid Mech., 119*, 249–267.

Chen, F. 1991. Throughflow effects on convective instability in superposed fluid and porous layers. *J. Fluid Mech., 231*, 113–133.

Chen, F. and Chen, C. F. 1988. Onset of finger convection in a horizontal porous layer underlying a fluid layer. *Trans. ASME C: J. Heat Transfer 110*, 403–409.

Chen, F. and Chen, C. F. 1989. Experimental investigation of convective stability in a superposed fluid and porous layer when heated from below. *J. Fluid Mech., 207*, 311–321.

Chouke, R. L., van Meurs, P. and van der Poel, C. 1959. The instability of slow, immiscible viscous liquid–liquid displacements in permeable media. *Trans. AIME, 216*, 188–194.

Combarnous, M. A. and Bories, S. A. 1975. Hydrothermal convection in saturated porous media. *Adv. Hydrosci., 10*, 231–307.

Combarnous, M. A. and LeFur, B. 1969. Transfert de chaleur par convection naturelle dans une couche poreuse horizontale. *C. R. Acad. Sci. Paris B, 269*, 1009–1012.

Compton, R. G. and Pritchard, K. L. 1990. The dissolution of calcite at pH > 7: kinetics and mechanism. *Phil Trans. R. Soc. Lond., 330*, 45–70.

Compton, R. G. and Unwin, P. R. 1990. The dissolution of calcite in aqueous solution at pH < 4: kinetics and mechanism. *Phil Trans. R. Soc. Lond., 330*, 1–45.

Cvetkovic, V. and Dagan, G. 1994. Transport of kinetically sorbing solute by steady random velocity in heterogeneous porous formations. *J. Fluid Mech., 265*, 189–215.

Dagan, G. 1982. Stochastic modeling of groundwater flow by unconditional and conditional probabilities, Part 1. *Water Resources Res., 18*, 813–833; Part 2, *Ibid, 18*, 835–848.

Dagan, G. 1984. Solute transport in heterogeneous porous formations. *J. Fluid Mech., 145*, 151–177.

Dagan, G. 1988. Time-dependent microdispersion for solute transport in anisotropic heterogeneous aquifers. *Water Resources Res., 24*, 1491–1500.

Daly, R. A. 1951. Elastic properties of materials of the earth's crust. In *Internal Constitution of the Earth*, ed. B. Gutenberg, pp. 50–86. New York: Dover.

Darcy, H. 1856. *Les fontaines publiques de la ville de Dijon*, pp. 647 + Atlas. Dalmont, Paris.

Davis, S. H., Rosenblat, S., Wood, J. R. and Hewett, T. A. 1985. Convective fluid flow and diagenetic patterns in domed sheets. *Amer. J. Sci., 285*, 207–223.

Drever, J. I. 1982. *The Geochemistry of Natural Waters*. Englewood Cliffs, NJ: Prentice-Hall.

Dunkle, S. A., Plummer, L. N., Busenberg, E., Phillips, P. J., Denver, J. M., Hamilton, P. A., Michel, R. L. and Coplen, T. B. 1993. Chlorofluorocarbons ($CCl_3 F$ and $CCl_2 F_2$) as dating tools and hydrologic tracers in shallow groundwater of the Delmarva Peninsula, Atlantic Coastal Plain, United States. *Water Resources Res., 29*, 3837–3860.

Dupuit, J. 1863. *Etudes théoretiques et practiques sur le mouvement des eaux dans les canaux découverts et à travers les terraines perméables*. Paris: Dunrod.

Dykhuizen, R. C. 1992. Diffusive matrix fracture coupling including the effects of flow channeling. *Water Resources Res., 28*, 2447–2550.

Ekmekci, M., Gunay, G. and Simsek, S. 1995. Morphology of the rimstone pools at Pammukkale, Western Turkey. *Cave and Karst Sci., 22*, 93–130.

Elder, J. W. 1965. Physical processes in geothermal areas. In *Terrestrial Heat Flow*, ed. W. K. H. Lee, pp. 211–239. Amer. Geophys. Union, monogr. ser. 8, Washington, D.C.

Elder, J. W. 1967. Steady free convection in a porous medium heated from below. *J. Fluid Mech., 27*, 29–48.

Engesgaard, P. and Kipp, K. L. 1992. A geochemical transport model for Redox-controlled movement of mineral fronts in groundwater flow systems: a case of nitrate removal by oxidation of Pyrite. *Water Resources Res., 28,* 2829–2843.

Evans, D. G. and Nunn, J. A. 1989. Free thermohaline convection in sediments surrounding a water column. *J. Geophys. Res., 94,* 12 413–12 422.

Ferry, J. M. 1987. Metamorphic hydrology at 13 km depth and 400–550°C. *Amer. Min., 72,* 39–58.

Ferry, J. M. 1991. A model for coupled fluid flow and mixed volatile mineral reactions with application to regional metamorphism. *Contrib. Mineral. Petrol., 106,* 273–285.

Ferry, J. M. and Dipple, G. M. 1992a. Metasomatism and fluid flow in ductile fault zones, *Contrib. Mineral. Petrol., 112,* 149–164.

Ferry, J. M. and Dipple, G. M. 1992b. Models for coupled fluid flow, mineral reaction and isotopic alteration during contact metamorphism. *Am. Mineral., 77,* 577–591.

Fitzgerald, S. D. and Woods, A. W. 1994. The instability of a vaporization front in hot, porous rock. *Nature, 367,* 450–453.

Freeze, R. A. and Cherry, J. A. 1979. *Groundwater.* New York: Prentice-Hall.

Freyberg, D. L. 1986. A gradient experiment on solute transport in a sand aquifer, Part 2. Spatial moments and the advection and dispersion of non-reactive tracers. *Water Resources Res., 22,* 2031–2046.

Gaetani, M. E., Fois, E., Jadoul, F. and Hicora, A. 1981. Nature and evolution of middle Triassic carbonate buildups in the Dolomites, Italy. *Marine Geol., 74,* 25–57.

Garabedian, S. P., LeBlanc, D. R., Gelhar, L. W. and Celia, M. A. 1991. Large-scale natural gradient tracer test in sand and gravel, Cape Cod, Massachusetts. 2. Analysis of spatial moments for a non-reactive tracer. *Water Resources Res., 27,* 911–924.

Garrels, R. M. and Christ, C. L. 1965. *Solutions, Minerals and Equilibria.* New York: Harper & Row.

Garrels, R. M. and Thompson, M. E. 1962. A chemical model for seawater at 25 °C and one atmosphere total pressure. *Amer. J. Sci., 260,* 57–66.

Gelhar, L. W. 1986. Stochastic subsurface hydrology from theory to applications. *Water Resources Res., 22,* 134S–145S.

Gelhar, L. W. and Axness, C. L. 1983. Three-dimensional stochastic analysis of macrodispersion in a stratified aquifer. *Water Resources Res., 19*(1), 161–180.

Gill, A. E. 1969. A proof that convection in a vertical slab is stable. *J. Fluid Mech., 35,* 545–548.

Goldhammer, R. K. 1987. Platform carbonate cycles, middle Triassic of northern Italy. Ph.D. dissertation, The Johns Hopkins University.

Goldhammer, R. K., Dunn, P. A. and Hardie, L. A. 1987. High frequency glacio-eustatic sea level oscillations with Milankovitch characteristics recorded in Middle Triassic platform carbonates in northern Italy. *Amer. J. Sci., 287,* 853–892.

Griffiths, R. W. 1981. Layered double-diffusive convection in porous media. *J. Fluid Mech., 102,* 221–248.

Habermann, B. 1960. The efficiency of miscible displacement as a function of mobility ratio. *Trans. AIME, 219,* 264–272.

Hanor, J. S. 1987. Kilometre-scale thermohaline overturn of pore waters in the Louisiana Gulf Coast. *Nature, 327,* June 11, p. 501.

Hardie, L. A., Bosellini, A. and Goldhammer, R. K. 1986. Repeated subaerial exposure of subtidal carbonate platforms, Triassic, northern Italy: evidence for high frequency sea level oscillations on a 10^4 year scale. *Paleooceanography, 1,* 447–457.

Harned, H. S. and Owen, B. B. 1958. *The Physical Chemistry of Electrolytic Solutions.* New York: Reinhold.

Hartline, B. K. and Lister, C. R. B. 1977. Thermal convection in a Hele-Shaw cell. *J. Fluid Mech., 79,* 379–389.

Haszeldine, R. S., Samson, I. M. and Cornford, C. 1984. Quartz diagenesis and convective fluid movement: Beatrice Oilfield, UK North Sea. *Clay Minerals, 19,* 391–402.

Hele-Shaw, H. J. S. 1898. The flow of water. *Nature, 58,* 34–36. *See also*: Mathematics at the British Association. *Ibid., 58,* 534–535.

Helgeson, H. C., Murphy, W. M. and Aagaard, P. 1984. Thermodynamic and kinetic constraints on reaction rates among minerals and aqueous solutions. II. Rate constants, effective surface area and the hydrolysis of feldspar. *Geochim. Cosmochim. Acta, 48,* 2405–2432.

Hess, K. M., Wolf, S. H. and Celia, M. A. 1992. Large-scale natural gradient tracer test in sand and gravel, Cape Cod, Massachusetts. Part 3. Hydraulic conductivity variability and calculated micro-dispersivities. *Water Resources Res., 28,* 2011–2027.

Hewett, T. A. 1986. Porosity and mineral alteration by fluid flow through a temperature field. In *Reservoir Characterization*, ed. L. W. Lake and H. B. Carroll, Jr., pp. 83–94. Orlando, FL: Academic Press.

Hill, S. 1952. Channeling in packed columns. *Chem. Eng. Sci., 1,* 247–253.

Hinch, E. J. and Bhatt, B. S. 1990. Stability of an acid front moving through porous rock. *J. Fluid Mech., 212,* 279–288.

Hogg, N. G., Katz, E. J. and Sanford, T. B. 1978. Eddies, islands and mixing. *J. Geophys. Res., 83,* 2921–2938.

Homsy, G. M. 1987. Viscous fingering in porous media. *Ann. Rev. Fluid Mech., 19,* 271–311.

Horton, C. W. and Rogers, F. T. Jr. 1945. Convection currents in a porous medium. *J. Appl. Phys., 16,* 367–370.

Hubbert, M. K. 1940. The theory of ground-water motion. *J. Geol., 48,* 785–944.

Hubbert, M. K. 1967. Application of hydrodynamics to oil exploration. *Proc. 78th World Petrol. Congress, 1B,* 59–75.

Huppert, H. E. and Woods, A. W. 1995. Gravity-driven flows in porous layers. *J. Fluid Mech., 292,* 55–69.

Imhoff, P. T. and Green, T. 1988. Experimental Investigation of double-diffusive groundwater fingers. *J. Fluid Mech., 188,* 363–382.

Jensen, K. H., Bitsch, K. and Bjerg, P. L. 1993. Large-scale dispersion experiments in a sandy aquifer in Denmark: Observed tracer movements and numerical analyses. *Water Resources Res., 29,* 673–696.

Joseph, D. D. 1976. *Stability of Fluid Motions.* New York: Springer-Verlag.

Kampé de Fériet, J. 1939. Les functions aléatoires stationnaires et la théorie statistique de la turbulence homogène. *Ann. Soc. Sci, Brux., 59,* 145.

Katto, Y. and Masuoka, T. 1967. Criterion for onset of convection flow in a fluid in a porous medium. *Int. J. Heat Mass Transfer, 10,* 297–309.

Kestin, J., Khalifa, H. E. and Correia, R. J. 1981. Tables of the dynamic and kinematic viscosity of aqueous NaCl solutions in the temperature range of 20°–150°C and the pressure range 0.1–35 MPa. *J. Phys. Chem. Ref. Data, 10,* 71–87.

Kohout, F. A. 1965. A hypothesis concerning cyclic flow of salt water related to geothermal heating in the Floridian aquifer. *Trans. N. Y. Acad. Sci., 28,* 249–271.

Kohout, F. A., Henry, H. R. and Banks, J. E. 1977. Hydrology related to geothermal conditions of the Florida Plateau. In *The Geothermal Nature of the Florida Plateau*, eds. D. L. Smith and G. M. Griffin. Fla. Bur. Geol., Spec. Publ. 21, pp. 1–41. Tallahassee, Fla.

Kvernvold, O. and Tyvand, P. A. 1980. Dispersion effects on thermal convection in porous media. *J. Fluid Mech., 99*, 673–686.

Lamb, H. 1932. *Hydrodynamics*. Cambridge University Press.

Lapwood, E. R. 1948. Convection of a fluid in a porous medium. *Proc. Camb. Phil. Soc., 44*, 508.

LeBlanc, D. R., Garabedian, S. P., Hess, K. M., Gelhar, L. W., Quadri, R. D., Stollenwerk. K. G. and Wood, W. W. 1991. Large-scale natural gradient tracer test in sand and gravel, Cape Cod, Massachusetts, Part 1. Experimental design and observed tracer movement. *Water Resources Res., 27*, 895–910.

Lerman, A. 1979. *Geochemical Processes: Water and Sediment Environments*. New York: Wiley.

Leverett, M. C. 1939. Flow of oil–water mixtures through unconsolidated sands. *Trans. Amer. Inst. Mech. Eng., 142*, 152–169.

Lichtner, P. C. 1985. Continuum model for simultaneous chemical reactions and mass transport in hydrothermal systems. *Geochim. Cosmochim. Acta, 49*, 779–800.

Lichtner, P. C. 1988. The quasi-stationary state approximation to coupled mass transport and fluid–rock interaction in a porous medium. *Geochim. Cosmochim. Acta, 51*, 143–165.

Lichtner, P. C. 1991. The quasi-stationary state approximation to fluid-rock interaction: local equilibrium revisited. *Adv. Phys. Geochem., 8*, 452–460.

Lichtner. P. C. 1992. Time-space continuum description of fluid/rock interaction in permeable media. *Water Resources Res., 28*, 3135–3155.

Lighthill, James 1696. *An Informal Introduction to Theoretical Fluid Mechanics*. Oxford University Press.

Lovley, D. R., Baedecker, M. J., Lonergan, D. J., Cozzarelli, I. M., Phillips, E. J. P. and Siegel, D. I. 1989. Oxidation of aromatic contaminants coupled to microbial iron reduction. *Nature, 339*, 297–299.

Lyle, S., Huppert, H. E., Hallworth, M., Bickle, M. and Chadwick, A. 2005. Axisymmetric gravity currents in a porous medium. *J. Fluid Mech., 543*, 293–302.

Mackay, D. M., Freyberg, P. V., Roberts, P. V. and Cherry, J. A. 1986. A natural gradient experiment on solute transport in a sand aquifer, Part 1. Approach and overview of plume movement. *Water Resources Res. 22*, 2017–2029.

Masterson, J. P., Hess, K. M., Walter, D. A. and LeBlanc, D. R. 2002. Simulated changes in the sources of ground water for public-supply wells, ponds, streams and coastal areas on Western Cape, Massachusetts. *Water Res. Investig. Rep., 02–4143*, U.S. Geol. Survey, Denver, CO.

McBride, E. F. 1989. Quartz cement in sandstones: A review. *Earth-Sci Rev., 26*, 69–112.

McManus, K. M. and Hanor, J. S. 1988. Calcite and iron sulfide cementation of Miocene sediments flanking the west Hackberry salt dome, southwest Louisiana, USA. *Chem. Geol., 74*, 99–112.

Menard, T. and Woods, A. W. 2005. Dispersion, scale and time dependence of mixing zones under gravitationally stable and unstable displacements in porous media. *Water Resources Res., 41*, W05014, doi:1029/2004WR003701.

Merino, E. 1984. Survey of geochemical self-patterning phenomena. In *Chemical Instabilities*, eds. G. Nicolis and F. Baras, pp. 305–328. Dordrecht: Reidel.

Merino, E., Ortoleva, P. and Strickholm, P. 1983. Generation of evenly spaced pressure-solution seams during late diagenesis: a kinetic theory. *Contrib. Mineral. Petrol., 82*, 360–370.

Monin, A. S. and Yaglom, A. M. 1975. *Statistical Fluid Dynamics*, Vol. 2, ed. J. L. Lumley. MIT Press.

Mulligan, Ann and Uchupi, E. 2003. New interpretation of glacial history of Cape Cod may have important implications for groundwater contamination transport. *Eos, Transactions, Amer. Geophys. Union, 84*, 177.

Murray, B. T. and Chen, C. F. 1989. Double-diffusive convection in a porous medium. *J. Fluid Mech., 201*, 147–166.

Muscat, M. and Meres, M. W. 1936. The flow of heterogeneous fluids through porous media. *Physics, 7*, 346–363.

Newman, D. K. 2003. Microbial mineral respiration. *The Bridge* (US Nat. Acad. Eng.), *33*, 9–14.

Nield, D. A. 1968. Onset of thermohaline convection in a porous medium. *Water Resources Res., 4*, 553–560.

Nield, D. A. 1984. Non-Darcy effects in convection in a saturated porous medium, pp. 129–140 in *Convective flows in porous media*, eds. R. A. Wooding and I. White. DSIR Sci. Info. Publ. Centre, Wellington, N.Z.

Nield, D. A. 1987a. Convective instability in porous media with throughflow. *AIChE J., 33*, 1222–1224.

Nield, D. A. 1987b. Throughflow effects in the Rayleigh–Bénard convective instability problem. *J. Fluid Mech., 185*, 353–360.

Nield, D. A. and Bejan, A. 1999. *Convection in a Porous Medium.* New York: Springer.

Nimmo, J. R. and Mello, K. A. 1991. Centrifugal techniques for measuring saturated hydraulic conductivity. *Water Resources Res., 27*(6), 1263–1269.

Ortoleva, P., Chadham, J., Merino, E. and Sen, A. 1987a. Geochemical self-organization. II. Reactive-infiltration instability. *Amer. J. Sci., 287*, 1008–1040.

Ortoleva, P., Merino, E., Moore, C. and Chadham, J. 1987b. Geochemical self-organization. I. Reaction-transport feedbacks and modeling approach. *Amer. J. Sci., 287*, 979–1007.

Palciauskas, V. V. and Domenico, P. A. 1976. Solution chemistry, mass transfer and the approach to chemical equilibrium in porous carbonate rocks and sediments. *Geol. Soc. Amer. Bull., 87*, 207–214.

Palm, E., Weber, J. E. and Kvernvold, O. 1972. On steady convection in a porous medium. *J. Fluid Mech., 54*, 153–162.

Park, C.-W and Homsy, G. M.. 1984. Two phase displacement in Hele-Shaw cells: theory. *J. Fluid Mech., 139*, 291–308.

Park, C.-W. and Homsy, G. M. 1985. The instability of long fingers in Hele-Shaw flows. *Phys. Fluids, 28*, 1583–1585.

Philip, J. R. 1989. The scattering analog for infiltration in porous media. *Rev. Geophys., 27*(4), 431–446.

Phillips, O. M. 1970. On flows induced by diffusion in a stably stratified fluid. *Deep-Sea Res., 17*, 435–43.

Phillips, O. M. 1977. *The Dynamics of the Upper Ocean.* Cambridge University Press.

Phillips, O. M. 1990. Flow controlled reactions in rock fabrics. *J. Fluid Mech., 212*, 263–78.

Phillips, O. M. 1991. *Flow and Reactions in permeable rocks.* Cambridge University Press.

Phillips, O. M. 2003. Groundwater flow patterns in extensive shallow aquifers with gentle relief: theory and application to the Galena/Locust Grove region of eastern Maryland. *Water Resources Res., 39* (6), 1149, doi:10.1029/2001WR001261.

Philpotts, A. R. 1990. *Principles of Igneous and Metamorphic Petrology.* Englewood Cliffs, NJ: Prentice-Hall.

Plummer, L. N. and Wigley, T. M. L. 1976. The dissolution of calcite in CO_2 – saturated solutions at 25 °C and 1 atmosphere total pressure. *Geochim. Cosmochim. Acta, 40*, 191–202.

Preuss, K., Faybishenko, B. and Bodvarrson, G. S. 1997. Alternative concepts and approaches for modelling unsaturated flow and transport in fractured rocks. In *The Site-Scale Unsaturated Zone Model of Yucca Mountain, Nevada, for the viability Assessment*, eds. G. S. Bodvarrson, T. M. Bandurraga and Y. S. Yu, ch. 24, LBNL-40376, pp. 281–322. Berkeley, California: Lawrence Berkeley Natl. Lab.

Prieto, C. and Destouni, G. 2005. Quantifying hydrological and tidal influences on groundwater discharges into coastal waters. *Water Resources Res., 41*, W12427, doi:1029/2004-WR003920.

Pryor, W. A. 1973. Permeability–porosity patterns and variations in some Holocene sand bodies. *Amer. Assoc. Petrol. Geol. Bull., 57*, 162–189.

Rayleigh, Lord. 1916. On convection currents in a horizontal layer of fluid when the higher temperature is on the under side. *Phil. Mag.*, ser. 6, *32*, 529–46.

Reilly, T. E., Plummer, L. N., Phillips, P. J. and Busenberg, E. 1994. The use of simulation and multiple tracers to quantify groundwater flow in a shallow aquifer. *Water Resources Res., 30*, 421–433.

Rice, S. O. 1944. Mathematical analysis of random noise. *Bell Syst. Tech. J., 23*, 282, and *24*, 46. Reprinted in *Selected Papers on Noise and Stochastic Processes* (ed. N. Wax), 1954. New York: Dover.

Roberts, P. V., Goltz, M. N. and Mackay, D. M. 1986. A natural gradient experiment on solute transport in a sand aquifer, Part 3. Retardation estimates and mass balances for organic solutes. *Water Resources Res., 22*, 2047–2058.

Rowland, J. V. and Sibson, R. W. 2004. Structural controls on hydrothermal flow in a segmented rift system, Taupo Volcanic Zone, New Zealand. *Geofluids, 4*, 259–283.

Rudraiah, N. and Srimani, P. K. 1980. Finite amplitude cellular convection in a fluid-saturated porous layer. *Proc. R. Soc. Lond. A, 373*, 199–222.

Russell, A. J., Kaar, J. L. and Berberich, J. A. 2003. Using biotechnology to detect and counteract chemical weapons. *The Bridge* (US Nat. Acad. Eng.), *33*(4), 19–24.

Saffman, P. G. and Taylor, G. I. 1958. The penetration of a fluid into a porous medium or Hele–Shaw cell containing a more viscous liquid. *Proc. R. Soc. Lond. A, 245*, 312–329.

Salve, R. and Oldenburg, C. M. 2001. Water flow within a fault in altered nonwelded tuff. *Water Resources Res., 37*, 12, 3043–3056.

Salve, R., Oldenburg, C. M. and Wang, J. S. 2003. Fault-matrix interactions in nonwelded tuff of the Paintbrush Group at Yucca Mountain. *J. Containment Hydrol., 62–63*, 269–286.

Shedlock, R. J., Denver, J. M., Hayes, M. A., Hamilton, P. A., Koterba, M. T., Backman, L. J., Phillips, P. J. and Banks, W. S. L. 1999. Water-quality assessment of the Delmarva Peninsula, Delaware, Maryland and Virginia: Results of investigations 1987–91. *U.S. Geol. Surv. Water Supply Pap., 2355-A*, 41pp.

Shercliff, J. A. 1975. Seepage flow in unconfined aquifers. *J. Fluid Mech., 71*, 181–192.

Sherwood, J. D. 1986. Island size distribution in stochastic simulation of the Saffman–Taylor instability. *J. Phys. A, 19*, 125–200.

Sibson, R. H. 1981. Fluid flow accompanying faulting. In *Fluid Flow Accompanying Faulting. Field Evidence and Models*, eds. D. W. Simpson and P. G. Richards. Earthquake prediction: an international review. Amer. Geophys. Union, Maurice Ewing Series 4, pp. 503–603.

Sibson, R. H. 1986. Brecciation processes in fault zones. *Pure and Appl. Geophys., 124,* 159–175.

Sibson, R. H. 1987. Earthquake rupturing as a mineralizing agent in hydrothermal syatems. *Geology, 15,* 701–704.

Sibson, R. H. 1989. Earthquake faulting as a structural process. *J. Struct. Geol., 11,* 1–14.

Sibson, R. H., Robert, F. and Poulsen, K. H. 1988. High-angle reverse faults, fluid-pressure cycling and mesothermal gold–quartz deposits. *Geology, 16,* 551–555.

Slobold, R. L. and Thomas, R. A. 1963. Effects of transverse diffusion on fingering in miscible-phase displacement. *Soc. Petrol. Eng. J., 3,* 9–13.

Sternlof, K. R., Karimi-Fard, M., Pollard, D. D. and Durlofsky, L. J. 2006. Flow and transport effects of compaction bands in sandstones at scales relevant to aquifer and reservoir management. *Water Resources Res., 42,* W07425, doi:10.1029/ 2005WR004664 2006.

Stober, I 1996. Researchers study conductivity of crystalline rock in proposed radioactive waste site. *Eos, Trans. Amer. Geophys. Union, 77, 10,* 93–94.

Sverdrup, H. U., Johnson, M. W. and Fleming, R. H. 1946. *The Oceans, their Physics, Chemistry and General Biology.* New York: Prentice Hall, Inc.

Sverjensky, D. A. 1986. Genesis of Mississippi Valley-type lead–zinc deposits. *Ann. Rev. Earth Planet. Sci., 14,* 177–199.

Taylor, G. I. 1921. Diffusion by continuous movements. *Proc. London Math. Soc.,* Ser 2, 196–212.

Tchalenko, J. S. and Ambraseys, N. N. 1970. Structural analysis of the Dasht-e-Bayaz (Iran) earthquake fractures. *Geol. Soc. Amer. Bull., 81,* 41–60.

Tokunaga, T., Nakata, T., Mogi, K., Watanabe, M., Shimada, J., Zhang, J., Gamo, T., Taniguchi, M., Asai, K. and Matsui, Y. 2002. Detection of submarine fresh groundwater discharge and its relation to onshore groundwater flow system. *Eos, Trans. Amer. Geophys. Union, 83*(47), Fall Meet. Suppl., Abstract H21B-0802.

Toth, D. J. and Lerman A. 1977. Organic matter reactivity and sedimentation rates in the ocean. *Amer. J. Sci., 277,* 465–485.

Tsypkin, G. G. and Woods, A. W. 2004. Vapour extraction from a water saturated geothermal reservoir. *J. Fluid Mech., 506,* 315–330.

Tsypkin, G. G. and Woods, A. W. 2005. Precipitate formation in a porous rock through evaporation of saline water. *J. Fluid Mech., 537,* 35–53.

Turner, J. S. 1973. *Buoyancy Effects in Fluids.* Cambridge University Press.

Valla, D. and Huppert, H. E. 2006. Gravity currents in a porous medium at an inclined plane. *J. Fluid Mech., 555,* 353–362. doi:1017/80022112006009578.

Walter, D. A. and Masterson, J. P. 2003. Simulation of advective flow under steady-state and transient recharge conditions, Camp Edwards, Massachusetts Military Reservation, Cape Cod, Massachusetts. *U.S. Geol. Survey, Water-resources Investigations Report* 03-4053, 51pp.

Walther, J. V. and Helgeson, H. C. 1979. Calculation of the thermodynamic properties of aqueous silica and the solubility of quartz and its polymorphs at high pressures and temperatures. *Amer. J. Sci., 277,* 1315–1351.

Washburn, E. W. 1929. *International Critical Tables.* New York: McGraw-Hill.

Weast, R. C. (ed.) 1972. *Handbook of Chemistry and Physics,* 53rd edn. Cleveland, Ohio: CRC Press.

Wiener, N. 1930. Generalized harmonic analysis. *Acta Math., 55,* 117–258.

Wilson, A. M. 2005. Fresh and saline groundwater discharge into the ocean: A regional perspective. *Water Resources Res., 41,* W02016, doi:1029/2004WR003399.

Wilson, E. N. 1989. Dolomitization of the Triassic Latemar buildup, Dolomites, northern Italy. Ph.D. dissertation. The Johns Hopkins University.

Wilson, E. N., Hardie, L. A. and Phillips, O. M. 1990. Dolomitization front geometry, fluid flow patterns and the origin of massive dolomite: the triassic Lateman buildup, Northern Italy. *Amer. J. Sci., 290,* 741–796.

Wilson, J. C. and McBride, E. F. 1988. Compaction and porosity evolution of Pliocene sandstones. Ventura basin, California. *Amer. Assoc. Petrol. Geol. Bull., 72,* 664–681.

Wood, J. R. 1987a. A model for dolomitization by pore fluid flow, pp. 17–35 in *Physics and Chemistry of Porous Media.* eds. J. R. Banavar, J. Koplik and K. W. Winker. New York: Amer. Inst. Phys.

Wood, J. R. 1987b. Calculation of mass transfer coefficients for dolomitization models. *Appl. Geochem., 2,* 629–638.

Wood, J. R. and Hewett, T. A. 1982. Fluid convection and mass transfer in porous limestones: a theoretical model. *Geochim. Cosmochim. Acta, 46,* 1707–1713.

Wooding, R. A. 1960. Rayleigh instability of a thermal boundary layer in flow through a porous medium. *J. Fluid Mech., 9,* 183–192.

Wooding, R. A. 1962. The stability of an interface between miscible fluids in a porous medium. *ZAMP, 13,* 255–266.

Wooding, R. A. 1963. Convection in a saturated porous medium at large Rayleigh number or Péclet number. *J. Fluid Mech., 15,* 527.

Wooding, R. A. 1969. Growth of fingers at an unstable diffusing interface in a porous medium or Hele-Shaw cell. *J. Fluid Mech., 30,* 477–495.

Wooding, R. A. and Morel-Seytoux, H. J. 1976. Multiphase fluid flow through porous media. *Ann. Rev. Fluid Mech., 8,* 233–274.

Woods, A. W. and Mason, R. 2000. The dynamics of 2-layer gravity-driven flows in permeable media. *J. Fluid Mech.,* 421, 83–114.

Zachara, J. M. and Rodin, E. E. 1996. Microbial reduction of crystalline iron oxides; influence of oxide surface area on potential for cell growth. *Envir. Sci. and Technol., 30,* 1618–1638.

Index